Bacteriocins: Ecology and Evolution

M.A. Riley   M.A. Chavan (Eds.)

Bacteriocins
Ecology and Evolution

With 15 Figures, 4 in Color, and 11 Tables

Prof. Dr. Margaret A. Riley
Dr. Milind A. Chavan
University of Massachusetts
Biology Department
221 Morril Science Center III
Amherst, MA 01003
USA

Library of Congress Control Number: 2006930602

ISBN-10 3-540-36603-2 Springer Berlin Heidelberg New York
ISBN-13 978-3-540-36603-4 Springer Berlin Heidelberg New York

This work is subject to copyright. All rights are reserved, whether the whole or part of the material is concerned, specifically the rights of translation, reprinting, reuse of illustrations, recitation, broadcasting, reproduction on microfilm or in any other way, and storage in data banks. Duplication of this publication or parts thereof is permitted only under the provisions of the German Copyright Law of September 9, 1965, in its current version, and permissions for use must always be obtained from Springer-Verlag. Violations are liable for prosecution under the German Copyright Law.

Springer is a part of Springer Science+Business Media

springer.com
© Springer-Verlag Berlin Heidelberg 2007

The use of general descriptive names, registered names, trademarks, etc. in this publication does not imply, even in the absence of a specific statement, that such names are exempt from the relevant protective laws and regulations and therefore free for general use.

Editor: Dr. Christina Eckey, Heidelberg, Germany
Desk editor: Dr. Andrea Schlitzberger, Heidelberg, Germany
Cover design: WMXDesign GmbH, Heidelberg, Germany
Production and typesetting: SPi

Printed on acid-free paper   SPIN 11536871   149/3100   5 4 3 2 1 0

Contents

1 **Introduction** 1
   MARGARET A. RILEY AND MILIND A. CHAVAN

2 **The Diversity of Bacteriocins in Gram-Negative Bacteria** 5
   DAVID M. GORDON, ELIZABETH OLIVER AND JANE LITTLEFIELD-WYER
   Summary 5
   2.1 Introduction 5
   2.2 The Frequency of Bacteriocin Production 7
       2.2.1 Colicins 7
       2.2.2 Microcins 8
   2.3 Bacteriocin Diversity 9
       2.3.1 Colicins 9
       2.3.2 Microcins 10
   2.4 Multiple Bacteriocin Production 11
   2.5 Overview 12
   References 17

3 **Molecular Evolution of Bacteriocins in Gram-Negative Bacteria** 19
   MILIND A. CHAVAN AND MARGARET A. RILEY
   Summary 19
   3.1 Introduction 19
   3.2 Bacteriocins of Gram-Negative Bacteria 20
   3.3 Colicins and Colicin-like Bacteriocins 21
       3.3.1 Colicin Gene Organization 21
       3.3.2 Functional Domains in Colicin and CLB Proteins 23
   3.4 Models of Colicin Evolution 25
       3.4.1 Diversifying Selection 25
       3.4.2 Diversifying Recombination 27
       3.4.3 Evolution of Colicin-like Bacteriocins 27
   3.5 Evolution of Colicin Killing Domains 29
   3.6 Evolution of the Translocation and Receptor-Binding Domains 36
   3.7 Evolution of Colicin Regulatory Sequences 37
   3.8 Colicin D: A Possible Intermediate Between Pyocins
       and Colicins 38
   3.9 Conclusions 40
   References 41

4 The Diversity of Bacteriocins in Gram-Positive Bacteria ... 45
NICHOLAS C.K. HENG, PHILIP A. WESCOMBE, JEREMY P. BURTON, RALPH W. JACK AND JOHN R. TAGG

Summary ... 45
4.1 Introduction ... 45
    4.1.1 Bacteriocins: A Historical Perspective ... 45
    4.1.2 Bacteriocins of Gram-Positive Bacteria ... 46
    4.1.3 Why Produce Bacteriocins? ... 47
    4.1.4 Detection of Bacteriocins of Gram-Positive Bacteria ... 48
    4.1.5 Nomenclature of Bacteriocins of Gram-Positive Bacteria ... 49
    4.1.6 Classification of Bacteriocins of Gram-Positive Bacteria ... 50
4.2 Class I: The Lanthionine-Containing (Lantibiotic) Bacteriocins ... 53
    4.2.1 Type AI Lantibiotics ... 53
    4.2.2 Type AII Lantibiotics ... 58
    4.2.3 Type B (Globular) Lantibiotics ... 61
    4.2.4 Type C (Multi-Component) Lantibiotics ... 62
4.3 Class II: The Unmodified Peptide Bacteriocins ... 64
    4.3.1 Type IIa: The Pediocin-like Peptides ... 64
    4.3.2 Type IIb: Multi-Component Bacteriocins ... 66
    4.3.3 Type IIc: Miscellaneous Unmodified Bacteriocins ... 67
4.4 Class III: The Large (>10 kDa) Bacteriocins ... 74
    4.4.1 Type IIIa: The Bacteriolysins (Bacteriolytic Enzymes) ... 74
    4.4.2 Type IIIb: The Non-Lytic Bacteriocins ... 78
4.5 Class IV: The Cyclic Bacteriocins ... 79
    4.5.1 Enterocin AS-48 ... 81
    4.5.2 Gassericin A and Reutericin 6 ... 82
    4.5.3 Uberolysin ... 82
4.6 Concluding Remarks ... 83
References ... 83

5 Peptide and Protein Antibiotics from the Domain *Archaea*: Halocins and Sulfolobicins ... 93
RICHARD F. SHAND AND KATHRYN J. LEYVA

Summary ... 93
5.1 Introduction ... 93
5.2 Halocins ... 94
    5.2.1 The Ubiquity of Halocin Production ... 94
    5.2.2 The Role of Halocins in the Environment and the Inability to Detect Halocin Activity in Hypersaline Crystallizer Ponds ... 98
    5.2.3 Activity Spectra ... 99
    5.2.4 Common Features of Halocins ... 101
    5.2.5 Microhalocins (≤10 kDa) ... 102
    5.2.6 Protein Halocins (>10 kDa) ... 104
5.3 Biotechnology of Halocins ... 106
5.4 Sulfolobicins ... 106
References ... 107

6 The Ecological and Evolutionary Dynamics of Model Bacteriocin Communities 111
BENJAMIN KERR
Summary 111
6.1 Introduction 111
6.2 Dynamics in Two-Strain Communities: Getting over the Hump 113
6.3 Dynamics in Three-Strain Communities: Playing Rock–Paper–Scissors 119
6.4 Evolution in Three-Strain Communities: Survival of the Weakest 125
6.5 Dynamics with many Strains: Universal Chemical Warfare 128
6.6 Discussion 129
Appendix: Sensitivity is an ESS in the Well-Mixed RPS Game 131
References 132

7 Bacteriocins' Role in Bacterial Communication 135
OSNAT GILLOR
Summary 135
7.1 Introduction 135
7.2 Bacteriocin-Mediated Intercellular Communication 137
    7.2.1 Autoregulation of Class I Bacteriocins 138
    7.2.2 Quorum Sensing Regulation of Class II Bacteriocins 138
7.3 Bacteriocin-Coordinated Multicellular Communication 139
    7.3.1 Oral Biofilms 140
    7.3.2 Gastrointestinal Biofilms 141
7.4 Conclusions 142
References 143

Subject Index 147

List of Contributors

BURTON, JEREMY P.
BLIS Technologies Ltd., Dunedin 9054, New Zealand, e-mail: jeremy.burton@blis.co.nz

CHAVAN, MILIND A.
Department of Biology, University of Massachusetts Amherst, Amherst, MA 01003, USA, e-mail: mchavan@bio.umass.edu

GILLOR, OSNAT
Department of Environmental Hydrology & Microbiology, Zuckerberg Institute for Water Research, Jacob Blaustein Institute for Desert Research, Ben Gurion University of the Negev, Sde Boqer Campus 84990, Israel, e-mail: gilloro@bgu.ac.il

GORDON, DAVID M.
School of Botany and Zoology, The Australian National University, Canberra, ACT 0200, Australia, e-mail: David.Gordon@anu.edu.au

HENG, NICHOLAS C.K.
Department of Microbiology and Immunology, Otago School of Medical Sciences, University of Otago, P.O. Box 56, Dunedin 9054, New Zealand, e-mail: nicholas.heng@stonebow.otago.ac.nz

JACK, RALPH W.
Department of Microbiology and Immunology, Otago School of Medical Sciences, University of Otago, P.O. Box 56, Dunedin 9054, New Zealand, e-mail: ralph.jack@stonebow.otago.ac.nz

KERR, BENJAMIN
Department of Biology, University of Washington, Box 351800, Seattle, WA 98195-1800, USA, e-mail: kerrb@u.washington.edu

LEYVA, KATHRYN J.
Department of Microbiology, Arizona College of Osteopathic Medicine, Midwestern University, Glendale, AZ 85308, USA, e-mail: kleyva@midwestern.edu

LITTLEFIELD-WYER, JANE
School of Botany and Zoology, The Australian National University, Canberra, ACT 0200, Australia, e-mail: Jane.Wyer@anu.edu.au

OLIVER, ELIZABETH
School of Botany and Zoology, The Australian National University, Canberra, ACT 0200, Australia, e-mail: Elizabeth.Oliver@anu.edu.au

RILEY, MARGARET A.
Department of Biology, University of Massachusetts Amherst, Amherst, MA 01003, USA, e-mail: riley@bio.umass.edu

SHAND, RICHARD F.
Department of Biological Sciences, Northern Arizona University, Flagstaff, AZ 86011, USA, e-mail: richard.shand@nau.edu

TAGG, JOHN R.
Department of Microbiology and Immunology, Otago School of Medical Sciences, University of Otago, P.O. Box 56, Dunedin 9054, New Zealand, e-mail: john.tagg@stonebow.otago.ac.nz

WESCOMBE, PHILIP A.
BLIS Technologies Ltd., Dunedin 9054, New Zealand, e-mail: philip.wescombe@blis.co.nz

1 Introduction

MARGARET A. RILEY AND MILIND A. CHAVAN

Microbes produce an extraordinary array of microbial defense systems. These include broad-spectrum classical antibiotics, metabolic byproducts, such as the lactic acids produced by lactobacilli, lytic agents such as lysozymes, numerous types of protein exotoxins, and bacteriocins, which are loosely defined as biologically active protein moieties with a bacteriocidal mode of action. This biological arsenal is striking not only in its diversity, but also in its natural abundance. Bacteriocins are found in almost every bacterial species examined to date, and within a species tens or even hundreds of different kinds of bacteriocins are present. Halobacteria universally produce their own version of bacteriocins, the halocins. Streptomycetes are characterized by broad-spectrum antibiotics. This diversity and abundance of antimicrobial weapons clearly suggest an important role for these potent antimicrobials. Less clear is how such diversity arose and what roles these biological weapons serve in microbial communities. One large family of antimicrobials, the protein-based bacteriocins, has served as a model for numerous, detailed explorations regarding their ecological roles and evolutionary histories. Bacteriocins differ from broad-spectrum, classical antibiotics in one critical way – they have a relatively narrow killing spectrum and are toxic only to bacteria closely related to the producing strain. These toxins have been found in all major lineages of Bacteria, and more recently, have been described as universally produced by some members of the Archaea. According to Klaenhammer, 99% of all bacteriocins may make at least one bacteriocin, and the only reason we have not isolated more is that few researchers have looked for them.

The bacteriocin family includes a diversity of proteins in terms of size, microbial targets, modes of action, and immunity mechanisms. The most extensively studied, the colicins produced by *Escherichia coli*, share certain key characteristics. Colicin gene clusters are encoded on plasmids and are composed of a colicin gene, which encodes the toxin; an immunity gene, which encodes a protein conferring specific immunity to the producer cell by binding to and inactivating the toxin protein; and a lysis gene, which encodes

Department of Biology, University of Massachusetts Amherst, Amherst, MA 01003, USA, e-mail: riley@bio.umass.edu; mchavan@bio.umass.edu

a protein involved in colicin release through lysis of the producer cell. Colicin production is mediated by the SOS regulon and is therefore principally produced under times of stress. Toxin production is lethal for the producing cell and any neighboring cells recognized by that colicin. A receptor domain in the colicin protein that binds a specific cell surface receptor determines target recognition. This mode of targeting results in the relatively narrow phylogenetic killing range often cited for bacteriocins. The killing functions range from pore formation in the cell membrane to nuclease activity against DNA, rRNA, and tRNA targets. Colicins, indeed all bacteriocins produced by Gram-negative bacteria, are large proteins. Pore-forming colicins range in size from 449 to 629 amino acids. Nuclease bacteriocins have an even broader size range, from 178 to 777 amino acids.

Although colicins are representative of Gram-negative bacteriocins, there are intriguing differences found within this subgroup of the bacteriocin family. *E. coli* encodes its colicins exclusively on plasmid replicons. The nuclease pyocins of *Pseudomonas aeruginosa*, which show sequence similarity to colicins, and other, as yet uncharacterized, bacteriocins are found exclusively on the chromosome. Other close relatives to the colicin family, the bacteriocins of *Serratia marcesens*, are found on both plasmids and chromosomes. In this volume, Chapter 2 further explores this fascinating abundance and diversity of bacteriocin proteins produced by Gram-negative bacteria, while Chapter 3 focuses on signatures of their evolutionary history contained within their DNA sequences.

Bacteriocins of Gram-positive bacteria are as abundant and even more diverse as those found in Gram-negative bacteria. They differ from Gram-negative bacteriocins in two fundamental ways. First, bacteriocin production is not necessarily the lethal event it is for Gram-negative bacteria. This critical difference is due to the transport mechanisms Gram-positive bacteria encode to release bacteriocin toxin. Some have evolved a bacteriocin-specific transport system, whereas others employ the *sec*-dependent export pathway. Second, Gram-positive bacteria have evolved bacteriocin-specific regulation, whereas bacteriocins of Gram-negative bacteria rely solely on host regulatory networks. The conventional wisdom about the killing range of Gram-positive bacteriocins is that they are restricted to killing other Gram-positive bacteria. The range of killing can vary significantly, from relatively narrow as in the case of lactococcins A, B, and M, which have been found to kill only *Lactococcus*, to extraordinarily broad. For instance, some types of lantibiotics, such as nisin and mutacin B-Ny266, have been shown to kill a wide range of organisms including *Actinomyces, Bacillus, Clostridium, Corynebacterium, Enterococcus, Gardnerella, Lactococcus, Listeria, Micrococcus, Mycobacterium, Propionibacterium, Streptococcus,* and *Staphylococcus*. Contrary to conventional wisdom, these particular bacteriocins are active also against a number of medically important Gram-negative bacteria including *Campylobacter, Haemophilus, Helicobacter,* and *Neisseria*. Chapter 4 provides a review of the diversity of Gram-positive bacteriocins.

The Archaea have their own distinct family of bacteriocin-like antimicrobials, known as archaeocins. The only characterized member is the halocin family produced by halobacteria, and few halocins have been described in detail. Archaeocins are produced as the cells enter stationary phase. When resources are limited, producing cells lyse sensitive cells and enrich the nutrient content of the local environment. As stable proteins, they may remain in the environment long enough to reduce competition during subsequent phases of nutrient flux. The stability of halocins may help explain why there is so little species diversity in the hypersaline environments frequented by halobacteria. Chapter 5 deals with peptide and protein antibiotics in the domain Archaea, focusing on halocins and sulfolobicins.

As is clear from this brief survey of bacteriocin diversity and distribution, this heterogeneous family of toxins is united only by the shared features of being protein-based toxins that are relatively narrow in killing spectrum, and often extremely hardy and stable. What makes these the weapons of choice in the microbial world remains an intriguing question. Chapters 6 and 7 provide compelling suggestions regarding the ecological role served by bacteriocins in microbial communities. As will be clear from these two reviews, we have only just begun to understand the fundamental roles these potent toxins serve.

Bacteriocins are now being explored for their potential utility in human and animal health applications, and agricultural uses. Before we will succeed in harnessing the vast power of these toxins to serve in human-mediated functions, we require a more complete understanding of how these proteins have evolved and are shared between bacterial lineages, and what roles they serve in natural microbial communities. Their application to meet numerous human challenges is limited only by our imagination and creativity. The microbial world has invested several billion years in selecting and refining the novel functions afforded by this heterogeneous class of proteins. Let's take advantage of this extraordinary evolutionary experiment!

2 The Diversity of Bacteriocins in Gram-Negative Bacteria

DAVID M. GORDON, ELIZABETH OLIVER AND JANE LITTLEFIELD-WYER

Summary

The frequency and diversity of bacteriocin production varies greatly among bacterial populations. The dynamic interactions occurring among bacteriocin-producing, sensitive and resistant cells are likely responsible for much of this variation. However, the frequency of bacteriocinogeny and the diversity of bacteriocins produced are also determined by the habitat in which the population lives and by the genomic background of the producing strains. The production by a cell of two or more bacteriocins is a common phenomenon, at least in *Escherichia coli*. Further research is required if we are to understand the nature of fitness advantages accruing from multiple bacteriocin production, and how to best exploit bacteriocins as replacements for traditional antibiotics and in the creation and selection of bacterial strains for use as probiotics.

2.1 Introduction

Allelopathy is the production of chemical compounds which are toxic to other organisms but not to the producers of these compounds. For microorganisms, there is a wealth of data demonstrating that allelopathy is an important mediator of intra- and inter-specific interactions and consequently, a significant factor in maintaining microbial biodiversity (Chap. 6, this volume). In bacteria, these allelopathic substances include metabolic by-products such as ammonia or hydrogen peroxide; the 'classical' antibiotics such as bacitracin and polymyxin B, lysozyme-like bacteriolytic enzymes and the bacteriocins.

The bacteriocins produced by Gram-negative bacteria are diverse. Over 30 bacteriocins from *Escherichia coli* have been identified and, undoubtedly, more have yet to be discovered. The diversity present in other Gram-negative species, including other members of the Enterobacteriaceae, is largely unexplored. The molecular mechanisms by which this diversity has arisen, at least

---

School of Botany and Zoology, The Australian National University, Canberra, ACT 0200, Australia, e-mail: David.Gordon@anu.edu.au; Elizabeth.Oliver@anu.edu.au; Jane.Wyer@anu.edu.au

Bacteriocins: Ecology and Evolution
(ed. by M.A. Riley and M.A Chavan)
© Springer-Verlag Berlin Heidelberg 2007

for that class of bacteriocins known as colicins, is well understood and is discussed elsewhere (Chap. 3, this volume). However, the factors influencing the frequency of bacteriocin production and the diversity of bacteriocins in populations of bacteria are largely unknown. Acquiring this knowledge is essential, not only if we are to understand the role bacteriocins play in shaping bacterial communities in natural environments but also because there is an increasing desire to exploit bacteriocins to solve a range of applied problems.

As a consequence of the rising incidence of resistance to most traditional antibiotics, numerous research programs have been implemented aiming to explore the potential role which naturally produced and genetically modified bacteriocins might have as replacements for traditional antibiotics (Gillor et al. 2004). Other efforts focus on the use of antimicrobial toxins as food preservatives (Gillor et al. 2004). There is also an ever-increasing interest in the use of bacteria as biocontrol agents for the management of fungal and bacterial plant pathogens and, more recently, as the active agent in probiotic formulations. Probiotic therapy is a disease prevention strategy used in humans and domesticated animals, as well as a procedure considered to enhance the growth rate of livestock and poultry. The basis of the method is to ensure the establishment of 'good' bacteria in the gastro-intestinal tract in order to prevent the establishment of bacterial pathogens. One of the most important attributes of a 'good' probiotic strain is thought to be the strain's ability to produce antimicrobial compounds. However, the successful use of bacteria as biocontrol agents will require a sound understanding of microbial ecology and the factors influencing the frequency of bacteriocin production and diversity in populations of bacteria. Thus, the aim of this chapter is not to describe the diversity of bacteriocins which have been characterised from Gram-negative bacteria but, rather, to identify and discuss some of the factors observed to influence bacteriocin diversity. As usual, much of the data relates to bacteriocin production in *E. coli* but reference will be made also to observations concerning other species of the Enterobacteriaceae. The material presented has largely not been published previously, and is based on phenotypic and genotypic surveys of bacteriocin production in three collections of *E. coli*. The first of these collections consists of 496 isolates taken from a variety of mammal species living throughout Australia. The methods used to isolate and characterise the strains in this collection have been described by Gordon and Cowling (2003). The second collection consists of 266 faecal isolates recovered from people living in the Canberra ACT region of Australia, the isolation and characterization of these strains having been described by Gordon et al. (2005). In addition to the human faecal isolates, the collection contains 353 isolates recovered from extra-intestinal body sites of people. The third dataset was collected and characterised using the same methods as those used for the other two, and consists of 208 strains recovered from soil, water and sediment samples from a variety of localities across Australia. All strains in the three collections have been screened using a PCR-based approach for the presence of 29 virulence-associated traits and to determine their *E. coli* group membership (A, B1, B2, D), as described by Gordon et al. (2005). The majority of the strains have been screened using a combination of phenotypic and genotypic approaches in order to determine the frequency of bacteriocin production and the types of bacteriocins present in these strains, using the methods described by Gordon and O'Brien (2006).

2.2 The Frequency of Bacteriocin Production

2.2.1 Colicins

Typically, a third of *E. coli* strains produces a mitomycin C-inducible bacteriocin (Riley and Gordon 1996). For example, 24% of the *E. coli* isolates from humans and 33% of the isolates from mammals were colicin producers. Although there have been relatively few representative surveys of other members of the Gram-negative bacteria, similar frequencies of production are observed in these studies (Reeves 1972; Riley et al. 2003). However, the prevalence of colicinogenic strains may vary from 10 to over 70% among different *E. coli* populations (Riley and Gordon 1996; Gordon et al. 1998). Much of the variation in the frequency of colicinogenic strains in populations of *E. coli* is undoubtedly due to the dynamic interactions occurring between colicin-producing, resistant and sensitive cells (Riley and Gordon 1999), which result in temporal fluctuations in the relative frequencies of these three phenotypes. Such temporal fluctuations have been observed in vitro (Kerr et al. 2002) and in a population of *E. coli* isolated from house mice, in which the frequency of colicinogenic isolates declined from 71 to 43% over a 7-month period (Gordon et al. 1998).

Although the frequency of colicinogenic strains in a population is expected to vary naturally, theoretical considerations and empirical observation suggest that the environment in which the cells live also influences the frequency of colicinogenic cells. The predictions of mathematical models indicate that colicin-producing cells will have an advantage in benign habitats (resulting in high population growth rates) whereas harsher habitats will favour non-producers (Frank 1994). There is some empirical evidence to support this prediction. Population growth rates and cell densities are significantly lower in the external environment than in the lower intestine of mammals and, in Australia, 9% of *E. coli* isolated from the environment were producers, a significantly lower proportion than the 30% observed for Australian faecal isolates. The frequency of colicin-producing *E. coli* also depends on the type of host from which the cells are isolated. In Australian mammalian carnivores, the frequency of colicin-producing strains is about 20% whereas colicin production is twice as frequent in strains isolated from herbivorous or omnivorous Australian mammals. For animals of similar body mass, the turnover rate of the carnivore gastro-intestinal tract is significantly faster than that recorded in herbivorous or omnivorous mammals (Hume 1999). The predictions of mathematical models support the observation that the cost of colicin production can result in colicinogenic strains being disadvantaged when living in hosts with high gut turnover rates (unpublished data). There is additional evidence for the importance of the host environment in determining the likelihood that a strain will be colicinogenic. There are two species in the genus *Hafnia* (Okada and Gordon 2003; Janda et al. 2005). A collection of *Hafnia* species 2 strains, isolated from fish, reptiles and mammals from across Australia, was screened for the presence of a bacteriocinogenic phenotype. The results of this screening showed that 4% of the isolates from fish, 64% of the isolates from reptiles and 29% of the isolates from mammals were bacteriocin producers.

Faecal isolates of *E. coli* can be assigned to one of four main genetic groups (subspecies), designated A, B1, B2 and D (Ochman and Selander 1984; Herzer et al. 1990). Strains of the four groups appear to occupy different ecological niches (Gordon and Cowling 2003; Gordon et al. 2005). For *E. coli*, it is well established that some traits, particularly virulence factors associated with extraintestinal disease, are largely confined to particular genetic groups (Johnson and Stell 2000). Genetic group membership also appears to predict the frequency of colicin production. Among the Australian mammal isolates, 46% of genetic group B2 strains produce a colicin, compared to only 23–27% of strains in the genetic groups A, B1 or D. The reasons for these differences are unknown.

2.2.2 Microcins

There has been far less work on the frequency of microcin production in *E. coli* or other Gram-negative bacteria. All the microcins characterised to date are secreted from the cell, rather than being released as a consequence of cell lysis (Braun et al. 2002). It has also been suggested that as much as 90% of the microcin produced by a cell may be retained within the cell (Braun et al. 2002). Consequently, there is no reliable and simple phenotypic method for assaying microcin production.

The *E. coli* isolated from Australian humans and mammals were screened for seven microcins, using a PCR-based approach. Of the human isolates screened, 32% were microcin producers whereas, among the isolates from mammals, 9% were microcin producers. Why microcin production is less common in *E. coli* isolated from mammals compared to humans is unknown. In the isolates from humans, microcin production is significantly more prevalent among genetic group B2 strains (47%) than among A (16%), B1 (18%) or D strains (9%). The frequency of microcin production in both of these *E. coli* collections has almost certainly been underestimated, as a PCR-screening approach can be used only for those microcins which have been genetically characterised.

2.3 Bacteriocin Diversity

2.3.1 Colicins

Surveys of colicin diversity in different collections of *E. coli* all give rise to similar results – only a small fraction of the known colicins are present in a given collection and, in general, different colicins are detected in different collections (Riley and Gordon 1996; Table 2.1). There are a few exceptions to this general trend – colicins E1 and Ia are often observed (Riley and Gordon 1996; Table 2.1). Colicin Ia is encoded on a conjugative plasmid and therefore is able to transfer among *E. coli* lineages, so the fact that it is one of the more commonly observed colicins could be expected. However, colicin B is also borne on a conjugative plasmid and is the most common colicin produced by the isolates from mammals, yet it is uncommon in human isolates (Table 2.1).

The available data also suggest that a single colicin type dominates in any particular population of *E. coli*. This observation is expected, based on our understanding of the dynamics of colicin-producing, resistant and sensitive cells. As the frequency of cells resistant to the dominant colicin type in a population increases, that of the dominant producer population will decline, thereby providing the opportunity for the invasion of a different type of producer to which the fraction resistant to the original colicin is susceptible. Thus, theory predicts that there should be a continual flux in the relative frequency of different colicin types in a population of *E. coli*. There is some

Table 2.1 The frequency of colicin types in three collections of *Escherichia coli* from Australia

Colicin type	Human isolates % Frequency	Mammal isolates % Frequency	Environmental isolates % Frequency
A	0	0	0
B	1.3	10.8	3.8
D	0.2	1.2	0
E1	8.9	3.6	-
E2	0.6	0	-
E6	0	0	-
E7	1.6	3.0	-
Ia	9.9	10.7	1.0
Ib	0.5	0	0
K	1.6	0	0
M	3.9	13.3	4.8
?[a]	9.7	25.0	7.2

[a]Identity of the colicin produced was not determined

evidence to support this prediction. Colicin E2 is rare among strains in the collection of *E. coli* recovered from Australian mammals (Table 2.1). However, in a study of *E. coli* isolated from a single Australian population of house mice, 25% of the colicinogenic strains produced colicin E2. Over a 7-month period, there was a significant decline in the frequency of E2 and a concomitant increase in the frequency of resistance to E2 (Gordon et al. 1998). Additional evidence comes from the distribution of colicin D. Overall, colicin D is rare in *E. coli* isolated in Australia (Table 2.1) but, among mammals, colicin D was detected only from a single population of mountain brush-tailed possum (*Trichosurus canis*).

The ecological niche of a bacterial strain may also play a role in determining bacteriocin diversity. Two new bacteriocins, alvecin A and B (Wertz and Riley 2004), have recently been described from genomic species 2 of the genus *Hafnia*. PCR screening for the presence of alvecins A and B revealed that these bacteriocins were most frequently produced by strains isolated from mammals but were not detected in bacteriocinogenic strains isolated from reptiles (unpublished data). The bacteriocinogenic isolates from reptiles appear to produce several novel, as yet uncharacterised bacteriocins which appear to be absent in isolates from mammals (unpublished data).

The frequency of bacteriocin production in *E. coli* varies depending on the genetic group membership of the producing strains, as does the type of bacteriocin produced by a strain. In the collection of isolates from mammals, colicin Ia is significantly more prevalent among group B2 strains (20%), less common in B1 strains (8%), uncommon in D strains (4%), and absent in group A strains. In the collection of isolates from humans, however, the frequency of colicin Ia producers is independent of a strain's *E. coli* group membership. In both the human and mammal *E. coli* collections, colicin E1 is significantly co-associated with K1, and a strain which is K1 positive is five times more likely to harbour the colicin E1 plasmid than if it is not.

2.3.2 Microcins

All of the seven microcins screened for were detected in *E. coli* isolated from humans and all, but one, were detected in the isolates from mammals (Table 2.2). As was the case for colicins, the distribution of a particular microcin varies based upon a strain's group membership. Microcin V is encoded on a conjugative plasmid and, in human isolates, its prevalence does not vary with a strain's *E. coli* group membership. However, V is not randomly distributed among *E. coli* genotypes. Thus, among the group B2 strains isolated from humans, microcin V is never detected in a strain encoding for either of the adhesins *focG* or *iha*, the toxin *hylA*, or the secreted protein *she*. Therefore, microcin V is absent from the 69% of the B2 strains which possess one or more of these traits but present in the 30% of the B2 strains which possess none of these traits.

Table 2.2 The frequency of microcin types in two collections of *E. coli* from Australia

Type	Human isolates % Frequency	Mammal isolates % Frequency
B17	1.8	3.8
C7	0.8	2.1
H47	21.5	3.8
J25	1.9	0
L	0.8	1.7
M	18.1	2.8
V	8.7	1.4

2.4 Multiple Bacteriocin Production

We have a good empirical and theoretical understanding of the dynamics of colicin-producing, resistant and sensitive cell populations (Chaps. 6 and 7, this volume). However, most research has focused on populations of producing cells encoding only for a single bacteriocin type, together with cells either resistant or sensitive to the bacteriocin being produced. Theoretical work by Czaran et al. (2002) investigated the dynamics of a community of cells where multiple toxin types were being produced, together with cells which had different sensitivity and resistance profiles to these toxins. Numerical simulations of the model revealed two distinct quasi-equilibriums, which Czaran et al. (2002) termed the "frozen" and "hyper-immunity" states. In the frozen state, all toxins are present in the community but most cells produce only a single toxin to which it is immune. In the hyper-immunity state, most cells produce no toxin, many others produce a single toxin, some produce multiple toxins, and a few produce most of the toxins present in the community. In the latter state, virtually all cells are resistant to most of the bacteriocins present in the population. Which outcome – frozen or hyper-immunity – eventuates depends on initial conditions, recombination rate, and the costs associated with toxin production. The results of the screening of the *E. coli* isolates from Australian human and mammals revealed that multiple bacteriocin production is common. In the human isolates, 35% of the bacteriocinogenic strains produced a single bacteriocin, 46% produced two, 18% produced three, and 1% produced four or more bacteriocins. The production of multiple bacteriocins is also common for strains isolated from mammals, where 52% of the bacteriocinogenic strains produced one type of bacteriocin, 30% produced two, 12% produced three, and 6% produced three or more different bacteriocins. Again, the genetic group membership of a strain influences the likelihood that a strain will produce multiple bacteriocins (Table 2.3). Among human isolates, group B2 strains are more likely to produce multiple

Table 2.3 The frequency of multiple bacteriocin production by an *E. coli* strain with respect to the strain's genetic group membership in two collections of *E. coli* from Australia

Collection	# of Bacteriocins	*E. coli* genetic group % of strains			
		A	B1	B2	D
Human isolates	1	57	60	26	54
	2	27	27	52	36
	3 or more	16	13	22	10
Mammal isolates	1	30	77	45	45
	2	60	17	29	40
	3 or more	10	6	26	15

bacteriocins compared to group A or B1 strains whereas, in the animal isolates, group A strains are most likely to be multiple producers. Over 40 different combinations of bacteriocins were observed among the *E. coli* strains isolated from Australian humans, and over 30 combinations from mammals (Table 2.4). The different bacteriocins observed in these two collections do not associate at random. A number of the bacteriocins co-occur at a significantly greater frequency than would be expected by chance; these include colicins Ia and E1, colicins B and M, microcins H47 and M, as well as colicin Ia and microcin V. Conversely, microcins H47 and V are significantly less likely to co-occur in a strain than would be expected by chance. One significant three-way association was detected: microcin J25 was most likely to be observed in a strain with colicin Ia and microcin V but not in strains encoding only Ia or V.

Of the combinations found to co-occur more frequently than expected by chance, all have the characteristic that at least one of the co-occurring bacteriocins is secreted by the cell, rather than released via cell lysis (Table 2.4). Few strains appeared to encode for two colicins which are also associated with lysis genes. Those that did – for example, colicins E2 and E7 – may represent chimeras. At least two examples of E2 and E7 chimeras have been reported which resulted from recombination of portions of the E2 and E7 colicin operons in a single plasmid and, in both cases, there is a single copy of the lysis gene (Tan and Riley 1997; Nandiwada et al. 2004). It may be that the co-occurrence of two colicins released via cell lysis imposes too high a cost to the cell due to the expression of two lysis genes.

2.5 Overview

For those colicins released via cell lysis, there is a wealth of mathematical theory (Levin 1988; Frank 1994; Durrett and Levin 1997) as well as in vitro (Chao and Levin 1981; Gordon and Riley 1999; Kerr et al. 2002) and

Table 2.4 Combinations of bacteriocins detected in a single strain of *E. coli* in two Australian collections

Human isolates Genotype	Frequency (%)	Mammal isolates Genotype	Frequency (%)
?[a]	4.6	?	21.1
B/cM[b]	1.1	B	0.6
B17	1.4	B/cM	19.3
B17/?	0.4	B/cM/B17/L	1.2
D/K	0.4	B/cM/C7	0.6
E1	4.9	B/cM/E1	1.2
E1/B/cM	0.4	B/cM/E1/H47/M	0.6
E1/B17	0.4	B/cM/E1/H47/M/L	0.6
E1/E2	0.4	B/cM/E7	0.6
E1/K	0.4	B/cM/H47/M	0.6
E1/cM	1.8	B/cM/H47/M/L	0.6
E1/cM/B17	0.4	B/cM/Ia	4.1
E2	0.4	B/cM/Ia/E1/E7	0.6
E7	2.5	B/cM/Ia/E7/C7	0.6
H47	5.3	B/cM/V	0.6
H47/E1	1.1	B17/?	0.6
H47/Ia	0.4	D	2.9
H47/Ia/V	0.4	D/C7	0.6
H47/J25/E1	0.4	E1	4.1
H47/L	0.4	E1/B17	0.6
H47/L/E1	0.4	E1/B17/L	0.6
H47/M	29.0	E7	4.7
H47/M/?	1.8	E7/H47/M	0.6
H47/M/B17	1.1	H47/M/?	1.8
H47/M/C7	1.1	Ia	16.4
H47/M/E1	3.2	Ia/B17	0.6
H47/M/E1/Ia/B17	0.4	Ia/C7/B17	0.6
H47/M/E7	0.7	Ia/E1	1.2
H47/M/Ia	0.4	Ia/E1/E7/C7/B17	0.6
H47/M/K	0.7	Ia/E7	0.6
H47/V	0.4	Ia/V	1.8
H47/V/E1	0.4	Ia/V/C7	0.6
Ia	4.9	V	0.6
Ia/B/cM	0.7	V/?	0.6

(Continued)

Table 2.4 Combinations of bacteriocins detected in a single strain of *E. coli* in two Australian collections—Continued

Human isolates Genotype	Frequency (%)	Mammal isolates Genotype	Frequency (%)
Ia/E1	2.1	cM	1.8
Ia/E1/cM	0.4	cM/?	1.2
Ia/V	5.3	cM/E1	0.6
Ia/V/C7	0.4	cM/E7	0.6
Ia/V/E1	1.8	cM/Ia	2.3
Ia/V/E2/E7	0.4	cM/Ia/B17	1.2
Ia/V/J25	2.5		
Ia/V/J25/E1	0.7		
Ia/V/K	0.7		
Ia/V/L	0.4		
Ib	0.4		
Ib/V	0.4		
Ib/cM	0.4		
J25	0.4		
K	1.1		
K/E2	0.4		
L	0.7		
M	0.7		
M/?	0.4		
M/C7	0.4		
V	4.6		
V/B/cM	0.4		
V/E1	0.4		
V/J25	0.4		
cM	3.2		

[a]The question mark denotes an unidentified colicin producer
[b]Abbreviations: cM denotes colicin M, and M denotes microcin M

in vivo (Kirkup and Riley 2004) experimental evidence which demonstrates the potential importance of colicins in mediating intra-specific interactions, and the highly dynamic nature of these interactions. However, our current understanding of the dynamics of bacteriocin production is restricted to those colicins released via cell lysis. Cell lysis represents a significant cost to the colicin-producing population, in addition to the costs associated with colicin plasmid carriage and colicin synthesis. The costs associated with colicin production are an important determinant of the fitness hierarchy among colicin-producing, resistant and sensitive cell populations (Riley and Gordon 1999; Kerr et al. 2002). This hierarchy is analogous to the game of Rock, Paper, Scissors, where producers out-compete sensitive cells, resistant cells out-compete producers, and sensitive cells out-compete resistant cells. Microcins are not released as a result of cell lysis, nor are a number of colicins such as B, Ia, Ib and M (Braun et al. 2002). The survey results suggest that the bacteriocins released through cell lysis represent a minority of the bacteriocins produced by *E. coli* (Riley and Gordon 1996). For example, of the bacteriocin-producing *E. coli* strains isolated from Australian humans, 71% do not encode for a bacteriocin released via lysis.

As suggested by Dykes and Hastings (1997), the fitness costs associated with producing a bacteriocin secreted from the cell may be significantly lower than the costs incurred by cells encoding for a bacteriocin released via cell lysis. If true, then one critical component of the Rock Paper Scissors scenario may be invalid for the secreted bacteriocins – that is, strains resistant to a secreted bacteriocin may not experience a universal fitness advantage when in competition with the strain producing a secreted bacteriocin. Although the resistant cells will be unaffected by the bacteriocin, the growth rate disadvantage which resistant cells suffer due to the loss or modification of an important surface receptor may be greater than the cost associated with producing a secreted bacteriocin. To date, no study has quantified the costs associated with producing a secreted bacteriocin. However, there is some evidence suggesting that the costs associated with producing any colicin are substantial, even if the colicin is not released due to cell lysis.

As described above, the frequency of colicin production is significantly less in *E. coli* strains isolated from mammalian carnivores compared to herbivorous or omnivorous mammals. For animals of a given body mass, gut turnover rates are significantly faster in mammalian carnivores, and a mathematical model predicts that, due to the costs associated with bacteriocin production, the fitness advantage accruing from bacteriocin production should decline as the turnover rate of the system increases. Colicins Ia and B represent the majority of the colicins produced by the isolates from Australian mammals. If any strain producing a colicin released via lysis is excluded from the analysis of the effect of diet on the frequency of colicin production, then the same result is obtained.

If, indeed, the lower frequency of colicinogenic strains in carnivorous mammals is due to the rapid turnover rate of the gastro-intestinal tract, then this would suggest that the costs of bacteriocin production without cell lysis are substantial. Evidently, experiments determining the costs associated with the production of secreted bacteriocins are required, as are experiments to determine if cells resistant to bacteriocins such as Ia or B out-compete strains producing these bacteriocins.

The likelihood that an *E. coli* strain will be bacteriocinogenic depends on the environment from which the host was isolated and on the genetic background of the strain, as does the type of bacteriocin present in a strain. The reasons for these patterns are poorly understood but it is clear that the magnitude of fitness advantage accruing from bacteriocin production is determined by many factors. The development of a successful probiotic strain will depend not only on the host to be targeted but also on the careful choice of the strain and the bacteriocin produced by that strain.

In *E. coli*, most bacteriocin-producing strains encode more than one type of bacteriocin. On the face of it, the advantage of multiple bacteriocin production is obvious. Consider a community initially consisting of a sensitive cell population and two populations of producing cells, each encoding a single bacteriocin type. Each of the producers can kill the other producer as well as sensitive cells. If one of the producing cells acquires, through recombination, the genes for the other colicin type, then this multiple colicin producer can kill sensitive cells and all cells encoding only a single colicin type. There are, however, no data demonstrating that a strain encoding, for example, colicin Ia and microcin V, will out-compete a strain producing only one of these bacteriocins.

There may also be other benefits arising from multiple bacteriocin carriage. In naturally occurring *E. coli* populations, resistance to colicins is a common phenomenon and most cells are resistant to most co-occurring colicins (Riley and Gordon 1992; Gordon et al. 1998; Feldgarden and Riley 1999). All of the E group colicins exploit the BtuB receptor, and a mutation which alters or causes the loss of this receptor will lead to resistance and render all or many E-type colicins ineffective (Feldgarden and Riley 1999). In those strains harbouring multiple bacteriocins, many (albeit not all) of the combinations represent bacteriocins which exploit different surface receptors. For example, the colicins Ia and E1 exploit the receptors Cir and BtuB. A mutation in one receptor is far more likely than the simultaneous occurrence of mutations in two different receptors. Thus, harbouring multiple bacteriocins exploiting different surface receptors may slow the evolution of resistance in populations where the dominant bacteriocinogenic strain produces multiple bacteriocins, compared to populations where the dominant bacteriocinogenic strain produces a single bacteriocin. Nevertheless, an expanded receptor repertoire cannot explain the occurrence of some of the co-associations observed to date. Microcins H47 and M are thought to exploit the same receptors (Cir Fiu IroN and FepA), whilst colicin Ia and microcin V are both thought to exploit the Cir receptor.

Microcin V production is induced when iron is limited. Although it is well established that iron is a limited resource in extra-intestinal body sites (Ratledge and Dover 2000), it is generally not considered to be limiting in the lower intestinal tract of vertebrates. By contrast, colicin Ia is induced under conditions of general nutrient limitation, a state which does occur in the lower intestinal tract. The ability to acquire iron when attempting to establish at extra-intestinal body sites is considered to be an important virulence trait in *E. coli* (Ratledge and Dover 2000). The strains responsible for an extra-intestinal infection are thought to originate from the *E. coli* community residing in the infected individual's intestinal tract (Mobley and Warren 1996). Therefore, strains causing extra-intestinal infections must have a suite of traits enabling them to invade and establish at extra-intestinal body sites as well as traits facilitating persistence in the intestine. Thus, we conjecture that the joint carriage of an SOS-induced colicin, such as Ia, and an iron-induced microcin (V) each confers a fitness advantage to the strain but in different environments: colicin Ia in the gut and microcin V at extra-intestinal body sites.

The bacteriocins produced by *E. coli* are diverse and this is undoubtedly true of other Gram-negative species, too. Many microcins and some colicins are difficult to detect using conventional phenotypic approaches and, given that overall 45% of the *E. coli* isolated from humans produce one or more detectable bacteriocins, it is conceivable that all strains of *E. coli* produce a bacteriocin. The survey results also illustrate that much of the diversity in bacteriocin production is a consequence of the association of multiple bacteriocins in a single cell. Understanding the adaptive significance of this diversity represents a considerable challenge but one which must be faced, if we are to successfully exploit bacteriocins in order to manage bacteria and the diseases they cause.

References

Braun V, Patzer S, Hantke K (2002) Ton-dependent colicins and microcins: modular design and evolution. Biochimie 84:365–380
Chao L, Levin BR (1981) Structured habitats and the evolution of anticompetitor toxins in bacteria. Proc Natl Acad Sci USA 78:6324–6328
Czaran TL, Hoekstra RF, Pagie L (2002) Chemical warfare between microbes promotes biodiversity. Proc Natl Acad Sci USA 99:786–790
Durrett R, Levin S (1997) Allelopathy in spatially distributed populations. J Theor Biol 185:165–171
Dykes GA, Hastings JW (1997) Selection and fitness in bacteriocin producing bacteria. Proc R Soc Lond B 264:683–687
Feldgarden M, Riley MA (1999) The phenotypic and fitness effects of colicin resistance in *Escherichia coli* K12. Evolution 53:1019–1027
Frank SA (1994) Spatial polymorphism of bacteriocins and other allelopathic traits. Evol Ecol 8:369–386
Gillor O, Kirkup BC, Riley MA (2004) Colicins and microcins: the next generation antimicrobials. Adv Appl Microbiol 54:129–146
Gordon DM, Cowling A (2003) The distribution and genetic structure of *Escherichia coli* in Australian vertebrates: host and geographic effects. Microbiology 149:3575–3586
Gordon DM, O'Brien CL (2006) Bacteriocin diversity and the frequency of multiple bacteriocin production in *Escherichia coli*. Microbiology (In Press)
Gordon DM, Riley MA (1999) A theoretical and empirical investigation of the invasion dynamics of colicinogeny. Microbiology 145:655–661
Gordon DM, Riley MA, Pinou T (1998) Temporal changes in the frequency of colicinogeny in *E. coli* from house mice. Microbiology 144:2233–2240

Gordon DM, Stern SE, Collignon PJ (2005) The influence of the age and sex of human hosts on the distribution of *Escherichia coli* ECOR groups and virulence traits. Microbiology 151:15-23

Herzer PJ, Inouye S, Inouye M, Whittam TS (1990) Phylogenetic distribution of branched RNA-linked multicopy single-stranded DNA among natural isolates of *Escherichia coli*. J Bacteriol 172:6175-6181

Hume ID (1999) Marsupial nutrition. Cambridge University Press, Cambridge

Janda JM, Abbott Sl, Bystrom S, Probert WS (2005) Identification of two distinct hybridization groups in the genus *Hafnia* by 16S rRNA gene sequencing and phenotypic methods. J Clin Microbiol 43:3320-3323

Johnson JR, Stell AL (2000) Extended virulence genotypes of *Escherichia coli* strains from patients with urosepsis in relation to phylogeny and host compromise. J Infect Dis 181:261-272

Kerr B, Riley MA, Feldman MW, Bohannan BJ (2002) Local dispersal promotes biodiversity in a real-life game of rock-paper-scissors. Nature 418:171-174

Kirkup BC, Riley MA (2004) Antibiotic-mediated antagonism leads to a bacterial game of rock-paper-scissors in vivo. Nature 428:412-414

Levin BR (1988) Frequency-dependent selection in bacterial populations. Philos Trans R Soc Lond B 319:459-472

Mobley HLT, Warren JW (1996) Molecular pathogenesis and clinical management. Urin Tract Infect 3:67-94

Nandiwada LS, Schamberger GP, Schafer HW, Diez-Gonzalez F (2004) Characterization of an E2-type colicin and its application to treat alfalfa seeds to reduce *Escherichia coli* O157:H7. Int J Food Microbiol 93:267-279

Ochman H, Selander RK (1984) Standard reference strains of *Escherichia coli* from natural populations. J Bacteriol 157:690-693

Okada S, Gordon DM (2003) Genetic and ecological structure of *Hafnia alvei* in Australia. Syst Appl Microbiol 26:585-594

Ratledge C, Dover LG (2000) Iron metabolism in pathogenic bacteria. Annu Rev Microbiol 54:881-941

Reeves P (1972) The bacteriocins. Springer, Berlin Heidelberg New York

Riley MA, Gordon DM (1992) A survey of Col plasmids in natural isolates of *Escherichia coli* and an investigation into the stability of Col plasmid lineages. J Gen Microbiol 138:1345-1352

Riley MA, Gordon DM (1996) The ecology and evolution of bacteriocins. J Indust Microbiol 17:151-158

Riley MA, Gordon DM (1999) A model of intraspecific microbial warfare. Trends Microbiol 7:129-133

Riley MA, Wertz JE (2002) Bacteriocins: evolution, ecology, and application. Annu Rev Microbiol 56:117-137

Riley MA, Goldstone CM, Wertz JE, Gordon DM (2003) A phylogenetic approach to assessing the targets of microbial warfare. J Evol Biol 16:690-697

Tan Y, Riley MA (1997) Nucleotide polymorphism in colicin E2 gene clusters: evidence for nonneutral evolution. Mol Biol Evol 14:666-673

Wertz JE, Riley MA (2004) Chimeric nature of two plasmids of *Hafnia alvei* encoding the bacteriocins alveicins A and B. J Bacteriol 186:1598-1605

3 Molecular Evolution of Bacteriocins in Gram-Negative Bacteria

MILIND A. CHAVAN AND MARGARET A. RILEY

Summary

The study of molecular evolution has become a valuable tool in understanding the origin of life and the speciation of organisms, with the focus on changes in DNA and protein sequence and their functions. Interest in studying the molecular evolution of bacteriocins, the narrow-spectrum peptide antimicrobials, was elicited due to the broad diversity and abundance of these proteins. The availability of a large amount of data on colicins, the bacteriocins produced by the Gram-negative bacterium, *Escherichia coli*, made it a model bacteriocin to study molecular evolution. Colicins have characteristic features which make them amenable resources in the investigation of the mechanisms employed in evolution. In this chapter, we have reviewed these features of colicins and we describe models proposed to explain how these antimicrobial proteins have evolved. Further, we have described how our current understanding of colicin evolution is important to the understanding of colicin-like bacteriocins produced by other Gram-negative bacteria.

3.1 Introduction

Microbes produce a remarkable array of toxins which help them to compete in their local environments for the limited niche space and nutritional resources available. The killing breadth of these toxins ranges from quite narrow (e.g., bacteriocins are potent antimicrobials which tend to target only members of the producing species) to quite broad (e.g., classical antibiotics often target highly divergent species of bacteria). Bacteriocins, a member of the narrow-spectrum toxins, have been described as the "microbial weapon of choice", based upon their abundance and diversity among producing bacteria (Reeves 1965; Riley and Wertz 2002b). By contrast, very few microbial

---

Department of Biology, University of Massachusetts Amherst, Amherst, MA 01003, USA, e-mail: riley@bio.umass.edu; mchavan@bio.umass.edu

Bacteriocins: Ecology and Evolution
(ed. by M.A. Riley and M.A Chavan)
© Springer-Verlag Berlin Heidelberg 2007

lineages produce broad-range toxins such as classical antibiotics (Berdy 1974; Hopwood and Chater 1980).

The first bacteriocin was reported in 1925 by Gratia, who observed inhibition of *Escherichia coli* ϕ by *E. coli* V. Since it killed *E. coli*, the inhibitory agent was initially referred to as "colicine" (renamed to colicin). Gratia and Fredericq generated a vast body of literature providing the very first, detailed information about the diverse family of colicin toxins (Fredericq 1957, 1963; Reeves 1965). It is largely the result of this impetus that the colicins became a model system for many subsequent studies of bacteriocin biochemistry, genetics, ecology and evolution (Riley and Gordon 1992; Riley 1993b; Braun et al. 1994; Gordon and Riley 1999; Smarda and Obdrzalek 2001; Riley and Wertz 2002a; Kirkup and Riley 2004). These studies have provided insights into the origins and function of colicins, their closest relatives (colicin-like bacteriocins produced by members of the Enterobacteriaceae), and certain types of pyocins, which are bacteriocins produced by *Pseudomonas* spp. (Parret and De Mot 2000; Riley et al. 2001; Michel-Briand and Baysse 2002). The primary focus of this chapter is to describe current knowledge regarding the evolution and diversification of colicins and their close relatives produced by Gram-negative bacteria.

3.2 Bacteriocins of Gram-Negative Bacteria

Gram-negative bacteria produce a wide variety of bacteriocins, which are specifically named after the genus (e.g., klebicins of *Klebsiella pneumoniae*) or species (e.g., colicins of *E. coli*, marcescins of *Serratia marcescens*, alveicins of *Hafnia alvei* and cloacins of *Enterobacter cloacae*) of the producing bacteria. The bacteriocins produced by *Pseudomonads* are generally referred to as pyocins.

This diversity of Gram-negative bacteriocins can be divided into three groups based on size: (1) large colicin-like (25–80 kDa) bacteriocins, (2) the much smaller microcins (<10 kDa) and (3) phage tail-like bacteriocins, which are multimeric peptide assemblies. Colicin-like large bacteriocins are SOS-inducible high molecular weight proteins (Braun et al. 1994). These bacteriocins will be described in following sections. Microcins are non-SOS-inducible low molecular weight peptides similar to the bacteriocins of Gram-positive bacteria (Baquero et al. 1978; Moreno et al. 2002). Phage tail-like bacteriocins are nuclease- and protease-resistant rod-like particles resembling a bacteriophage tail, which kill sensitive cells by depolarization of the cell membrane (Stachura et al. 1969; Kuroda and Kagiyama 1983; Nakayama et al. 2000; Strauch et al. 2003). These are proposed to be defective phages or to have originated from phages which evolved to function as bacteriocins. For example, pyocin R2 (produced by *Pseudomonas* spp.) appears to be a remnant of phage P2 whereas pyocin F2 is similar to phage lambda (Nakayama et al. 2000).

3.3 Colicins and Colicin-like Bacteriocins

Colicins are a diverse group of SOS-inducible high molecular weight bacteriocins produced by *Escherichia coli*. Colicin-encoding genes are found on plasmids, and numerous types of colicin molecules have been reported in the literature, as described in Chapter 2 (this volume). They are detected at high frequencies in natural populations of *E. coli* (Table 2.1 in Chap. 2), a feature indicative of their importance in microbial ecology (Riley and Gordon 1992).

Other members of the Enterobacteriaceae family also exhibit a high frequency (30–50%) of bacteriocin production (Pugsley 1984). Many of these bacteriocins are similar to colicins in structure and function, and share many molecular, evolutionary and ecological features as well. They are often referred to as colicin-like bacteriocins (CLBs). Historically, the interest in bacteriocins produced by, for example, *Klebsiella* sp., *Enterobacter* sp. and *Serratia* sp. has been limited to those which assist in the typing of clinical isolates. Some CLBs have been cloned and sequenced (Guasch et al. 1995; Riley et al. 2001; Wertz and Riley 2004; Chavan et al. 2005) but additional studies are required to begin to understand the diversity of CLBs. Similarly to colicins, CLBs have narrow killing spectra which are generally restricted to closely related species (Riley et al. 2003). Thus, klebicins, alveicins and marcescins have a killing range generally limited to *Klebsiella*, *Hafnia* and *Serratia* species respectively (Riley et al. 2003).

Bacteriocins produced by *Pseudomonas* spp. (pyocins) have also been extensively studied (Kuroda and Kagiyama 1983; Sano et al. 1993b; Duport et al. 1995). The pyocin genes are chromosomally encoded and are ubiquitous in *Pseudomonas*. Three types of pyocins have been described: F type, R type and S type. R- and F-type pyocins are produced by more than 90% and S-type by 70% of surveyed *P. aeruginosa* strains (Michel-Briand and Baysse 2002). Due to such high frequencies of these three types of pyocins, a *Pseudomonas* strain often produces more than one pyocin. The F and R types of pyocins are phage tail-like bacteriocins resistant to nucleases and proteases. The S-type pyocins are protease-sensitive bacteriocins similar to colicins and hence, comparison of S-type pyocins with colicins will be emphasized here.

3.3.1 Colicin Gene Organization

Colicin gene clusters consist of three tightly linked genes encoding the toxin, immunity and lysis proteins, and are usually found on plasmids. The toxin gene encodes the activity which kills the target cells. The immunity gene encodes a protein which protects the host cell from the killing action of its own colicin protein and from colicin produced by its clones (Fredericq 1957). Immunity protein binds adjacent to the active site of the colicin protein, and inhibits its activity by steric hindrance and electrostatic repulsion mechanisms. The lysis protein (also called the bacteriocin release protein) lyses the host cell to release the expressed bacteriocin proteins outside the cell. The lysis gene is sometimes absent, particularly when more than one colicin gene cluster coexist in the same cell. This interesting aspect of co-occurring colicins is discussed in Chapter 2 (this volume).

Colicins kill target cells by pore formation, nuclease activity or by disrupting the cell wall. Figure 3.1 presents the arrangement of genes encoding pore-former and nuclease colicins. The orientation of the immunity gene with reference to the colicin genes is different for nuclease and pore-forming colicin gene clusters. In the case of pore formers (e.g., colicins A, B, K, N, E1), the immunity gene is orientated opposite to the toxin gene. The immunity and toxin genes are co-linear in nuclease colicins (e.g., colicins E2-E9), thus forming an operon consisting of toxin, immunity and lysis genes. Some nuclease colicins also have additional genes providing immunity to additional colicins. For example, the colicin E3 gene cluster contains an additional gene encoding immunity to colicin E8 (Toba et al. 1988).

The gene clusters of pore-forming and nuclease colicin-like bacteriocins show similar organization as is observed for pore-forming and nuclease colicins respectively (Riley et al. 2001; Wertz and Riley 2004; Chavan et al. 2005). Immunity genes of S-type nuclease pyocins are expressed from the same strand as in nuclease colicin gene clusters whereas the immunity gene for pyocin S5, a pore former, is expressed from the opposite strand, similar to that of pore-forming colicins. A lysis gene has not been identified for pyocin gene clusters. However, Nakayama et al. (2000) suggest that the lytic systems described in *P. aeruginosa* PAO1 for R2 and F2 pyocins may be shared by S-type pyocins. In the case of 28B, a bacteriocin produced by *Serratia marcescens*, *regA* and *regB* genes encoding phage holin and phage lysozyme-like proteins are proposed to serve in the release of bacteriocins (Ferrer et al. 1996).

Colicin expression is regulated by the SOS induction system, which is mediated by the LexA repressor binding to an inverted repeat sequence

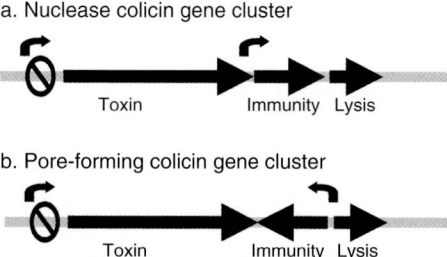

Fig. 3.1 Genetic organization of **a** nuclease and **b** pore-forming colicins. The orientation of genes encoding for killing activity, immunity and lysis proteins is depicted with *arrows*. The transcriptional starts are indicated by *curved arrows*. The LexA-binding region is indicated by ⊘

between the promoter and the ribosome binding site (Varley and Boulnois 1984). Some bacteriocins (28B and pyocins) do not have a LexA-binding box and are indirectly regulated by the SOS response via some other repressor protein (Matsui et al. 1993; Ferrer et al. 1996). The colicin and lysis genes share the same promoter whereas the immunity gene is induced from a constitutive promoter which maintains a certain basal level of immunity protein at all times to protect the producing cell.

When the SOS response is triggered in cells at times of stress, colicin genes are rapidly induced to express high levels of protein. In the case of nuclease colicins, the co-linear arrangement of the immunity and colicin genes within the gene cluster results in increased co-expression of the immunity protein which will bind to newly synthesized colicins and protect the cells from its nuclease activity. In the case of pore-forming colicins, induction does not result in increased levels of immunity protein, as the immunity gene is transcribed from the other strand. Pore-forming colicins, unlike the nuclease colicins, can kill the cells only from the exterior by punching holes in the cell membrane. Therefore, it may not be necessary for the cells to increase the levels of immunity protein during a phase of rapid colicin expression.

The colicin gene cluster consisting of toxin, immunity and lysis genes appears to have evolved to efficiently utilize the cellular resources for maximal colicin activity during colicin induction. The organization of toxin and immunity genes for pore forming ensures that resources are not wasted to express immunity protein after induction of pore-former colicin. However, cells expressing nuclease colicins are immediately protected by simultaneous induction of immunity proteins from the lethal enzymatic activity of their own colicins. This increased protection provides the cells sufficient time to produce enough colicins before their lysis.

3.3.2 Functional Domains in Colicin and CLB Proteins

The colicin and many CLB toxin proteins are organized into three functional domains: the N-terminal translocation, the central receptor binding, and the C-terminal killing domains (Fig. 3.2a). Pyocin proteins also contain these three domains but these are organized differently, and include an additional domain of unknown function (Fig. 3.2b). The receptor-binding and translocation domains of a colicin are required for uptake of colicin into sensitive cells. Together, these domains enable the colicin to recognize and enter target cells and thus, determine the specificity of colicin killing. The killing domain is responsible for the colicin activity which kills the sensitive cells by various mechanisms described below.

The interaction of a colicin molecule with the target cell is initiated by the binding of the receptor-binding domain to a specific cell surface receptor located on the outer cell surface. For example, colicins A and E1-E9 bind to the outer membrane protein BtuB, which is also known as the vitamin B12

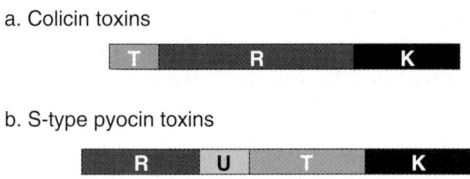

Fig. 3.2 Bacteriocin protein domain organization for **a** colicins and **b** S-type pyocins: translocation domain (*T*), receptor-binding domain (*R*), killing domain (*K*) and pyocin-specific domain (*U*), of as yet unknown function

receptor. The colicin protein is subsequently imported into the cell via the translocation domain utilizing either the TolA or TonB translocation system to move across the cell's outer membrane to reach the inner membrane (in the case of pore formers) or the cytoplasm (in the case of the nucleases). The killing domain then mediates the killing of a target cell by pore formation or nuclease activity. Pore formers (colicins A, B, K, N, S4, etc.) kill cells by creating pores in the cell membrane (Ohno-Iwashita and Imahori 1982). Nuclease colicins have DNase or RNase activities which degrade 16S rRNA or tRNAs. DNase colicins (colicins E2, E7, E8, E9) act non-specifically to digest target DNA (Toba et al. 1988; Chak et al. 1991). RNase colicins (colicins D, E3, E4, E5, E6) inactivate the protein biosynthetic machinery by targeting either 16S rRNA or tRNAs. Colicins E3, E4 and E6 have ribosomal endonuclease activity which hydrolyzes 16S rRNase (Bowman et al. 1971). Colicins D and E5 have hydrolytic activities on specific tRNA. Colicin D targets the four arginine isoacceptors (Tomita et al. 2000; Masaki and Ogawa 2002) whereas colicin E5 hydrolyses tRNAs for tyrosine, histidine, asparagine and aspartic acid, which contain a modified base, quenuine, at the wobble position of each anticodon (Ogawa et al. 1999). Additionally, a muraminidase function has been described for colicin M and pesticin activity. Muraminidase colicins degrade murein in the bacterial cell wall and thereby affect the cell's structural integrity, resulting in cell lysis (Schaller et al. 1982; Vollmer et al. 1997).

The S-type pyocins have a modular structure consisting of four domains. The S pyocin domains are, from N-terminal to C-terminal, a receptor-binding domain, a domain of unidentified function, a translocation domain, and a cytotoxic (killing) domain (Fig. 3.2b). Pyocin S1 lacks the second domain with unidentified function. As in the case of colicins, receptor-binding and translocation domains determine the species specificity whereas killing domains determine the mode of pyocin activity. A cognate immunity protein protects the pyocin-producing cell by binding to and masking the killing domain. However, further studies are required to define the receptor-binding and translocation domains of pyocin S5, which appear to have a different domain organization from that of the other S pyocins, as well as for colicin-like bacteriocins.

Initial studies indicated that pyocins S1 and S2 also inhibit lipid synthesis in sensitive cells, *P. aeruginosa* PML1516d, in addition to DNA damage (Sano et al. 1993b). Pyocin S3 did not exhibit such an effect on lipid metabolism in sensitive strain *P. aeruginosa* PA03092. Subsequent research with pyocin S1/S2/S3 chimeras indicated that none of the chimera pyocins caused inhibition of lipid synthesis in strain PA03092 but did so in strain PML1516d (Sano et al. 1993a). This lipid synthesis inhibition in strain PML1516d was not attributable to a particular domain. The inhibition of lipid synthesis is thus a species-specific secondary event in strain PML1516d, due to pyocin activities.

The domains of colicins and pyocins act in concert to kill sensitive cells. The receptor-binding and translocation domains exploit the host cellular infrastructure for the import of colicin into the cell. The activity of the cytotoxic domain then mediates the killing of the host/sensitive cell.

3.4 Models of Colicin Evolution

Riley and co-workers have extensively studied the evolution of colicins, and it is the result of these efforts that colicins are being recognized as a model system for the evolution of Gram-negative bacteriocins (Riley 1993a, 1993b, 1998; Riley et al. 1994, 2000; Tan and Riley 1996, 1997a; Riley and Wertz 2002a, 2002b). Two mechanisms – diversifying selection and recombination – have been proposed to explain how these molecules have evolved (Tan and Riley 1997b). According to the diversifying selection model, positive selection creates novel colicins by a series of point mutations which initially generate novel immunities and, subsequently, novel killing domains. The diversifying recombination model explains how diversity is created by the recombination of different colicin domains. Newly discovered colicin-like bacteriocins from related bacteria also exhibit characteristic features of molecular evolution proposed for colicins in prior studies (Riley et al. 2001; Wertz and Riley 2004; Chavan et al. 2005). Similarities observed in killing domains of colicins and CLBs indicate a common origin for these domains. Lateral transfer of genes or DNA segments may be responsible for the spread of killing domains.

3.4.1 Diversifying Selection

The colicin E3 and E6 gene clusters exhibit an interesting pattern of divergence. These two gene clusters have 95% identity within DNA sequences spanning their genes encoding for toxins and immunity proteins. However, the changes in nucleotide bases are not evenly distributed. Most mutations have accumulated within the region spanning the colicin-killing domain and the 5' half of the immunity gene. Thus, the immunity region and the immunity-binding region of the colicin gene exhibit high levels of polymorphism, which are not observed for the translocation and receptor-binding domains of colicins. A similar pattern was recorded in the divergence of colicins

E2 and E9. The role of positive selection was invoked, and a diversifying selection model was proposed to explain these patterns of nucleotide divergence in these two pairs of nuclease colicins (Riley 1993a, 1993b).

Riley and co-workers have proposed a "diversifying selection" hypothesis in which sequence diversification is thought to occur in two stages, as shown in Fig. 3.3. First, a point mutation occurs in the immunity gene, giving it a broadened immunity function (the ability to bind multiple colicins) and thereby conferring a selective advantage ensuring higher fitness than for its wild-type neighbours in a population producing multiple colicins. Second, a compensating mutation in the toxin gene of the colicin occurs in the immunity protein-binding site, thereby generating a "super killer" phenotype (Riley 1993b). Superkillers kill ancestral colicin-producing cells, since the immunity gene of the ancestral strain is no longer capable of binding the new toxin (Fig. 3.3d). At the same time, the newly evolved colicin-producing cell is protected from the ancestral colicin, since its immunity protein can still bind to the ancestral toxin.

The mechanism by which elevated mutation rates are achieved in colicin gene clusters may be linked to the physiology of the cell during SOS induction. During SOS, the expression of the error-prone DNA polymerase Pol IV is elevated, leading to an 800-fold increase in the mutation rate of replicated plasmids (Kim et al. 1997). Colicin plasmids may also mutate rapidly during this phase, leading to the evolution and selection of a new colicin gene cluster, as proposed in the diversifying selection model for colicin evolution.

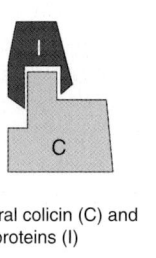

(a) Ancestral colicin (C) and immunity proteins (I)

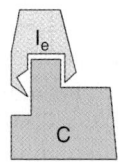

(b) Mutation(s) in immunity gene. Evolved immunity protein (I_e) still binds ancestral colicin

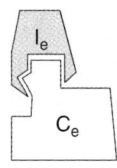

(c) Mutations in colicin gene. Selection of new colicin (C_e) that specifically binds to new immunity

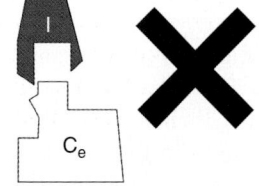

(d) Ancestral immunity can not bind to new colicin. Cells with ancestral colicin and immunity are no longer protected from new colicin.

Fig. 3.3 A model depicting diversifying selection for colicins

According to the diversifying selection model, a new toxin–immunity protein combination evolves by modification of the protein-binding interactions between the toxin and immunity proteins. Thus, new killing domains have evolved by generating new immunity specificity. The co-evolution of colicins and associated immunity proteins proposed in this model ensures that high-affinity binding interactions between the toxin and immunity proteins are selected so that the cells are not killed by their own toxins. This model of evolution is particularly well documented in nuclease colicins, of which the short killing domains (85–110 amino acid residues) have limited flexibility for diversification because mutations affecting the functional domain may render these colicins inactive.

3.4.2 Diversifying Recombination

All characterized, pore-former colicin proteins have high levels of localized sequence similarity to other pore-former colicins, creating a patchwork of shared and divergent sequences. The locations of the different patches frequently correspond to the different functional domains of the proteins. These observations led to the proposal that the diversity of colicins is increased by recombination events which exchange different domains (Braun et al. 1994; Tan and Riley 1997b). Such domain-based shuffling between bacteriocins is responsible for much of the variability observed among pore formers and also in nuclease colicins (e.g., colicins E6 and E7).

Examples of mix-and-match patterns of colicin domains are shown in Fig. 3.4. This is particular striking in colicins B and D, which have almost identical N-terminal translocation and receptor-binding domains. Their killing domains, however, have been selected from different sources. Colicin B has a colicin Y-like pore-forming domain whereas colicin D has a tRNase domain similar to that of klebicin D and colicin E4. Comparison of colicin 5, 10 and K sequences indicate the role of recombination events in creating new colicins by exchanging domains (Pilsl and Braun 1995). Colicin 5 has a colicin 10-like N-terminal region but a colicin K-like killing domain (Fig. 3.4). Colicins Ia and Ib also have identical N-terminal domains but their killing domains have only 51% identity.

Thus, there are two phases in colicin evolution. Initially, novel colicins are created by diversifying selection. If the novel colicins are successful, they become abundant and further diversities are subsequently created by recombination events (Riley and Wertz 2002b).

3.4.3 Evolution of Colicin-like Bacteriocins

Colicins and CLBs possess some common features: gene cluster organization, tri-domain organization of toxin, SOS-mediated expression, and shared

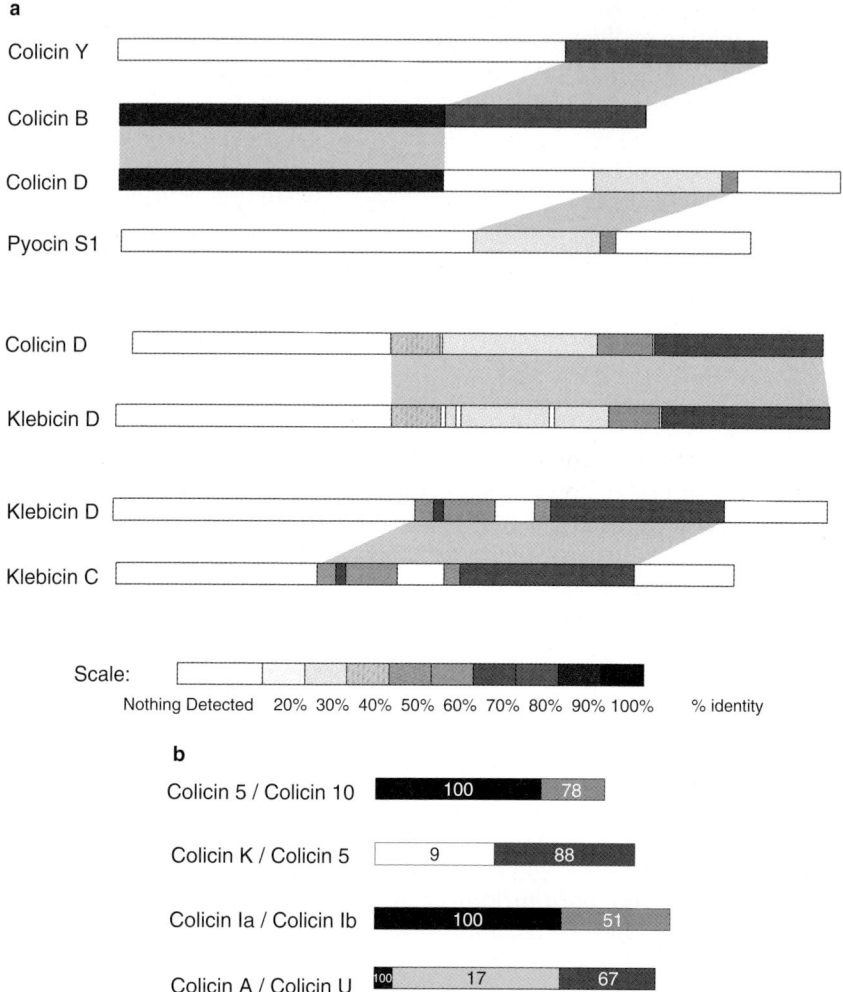

Fig. 3.4 A model depicting diversifying recombination for colicins. **a** Shared regions among colicins Y, B, D, pyocin S1, and klebicins C and D. The *scale* indicates the percentage identities between the shared regions. The *grey highlight* emphasizes the shared region between two individual activity proteins. **b** Pairs of colicins with regions of significant similarity indicated by % identity

killing domains. Riley and colleagues investigated whether or not the diversifying selection and recombination models developed to describe the molecular evolution of colicins could be extended to CLBs produced by other members of the Enterobacteriaceae family to which *E. coli* belongs. If applicable, the vast knowledge available in the colicin literature could be applied to bacteriocins produced by related species such as *Klebsiella* sp., *Enterobacter* sp., *Serratia* sp., *Hafnia* sp. and *Citrobacter* sp. Sequence data of bacteriocins cloned from *Klebsiella* and *Hafnia* sp. indicate that these are indeed similar to colicins in mode of action, protein structure, and genetic organization (Riley et al. 2001; Wertz and Riley 2004; Chavan et al. 2005).

The most recent illustration of diversifying recombination is seen from recently published sequences of klebicins (all of which have nuclease activity) and alveicins. The klebicin B gene cluster shares sequence similarity with colicin A in the 5' regulatory region and 3' lysis gene (Riley et al. 2001). Although the source of its translocation and receptor-binding domains are unknown, it has a nuclease colicin-like killing domain (most similar to pyocin S1) and immunity gene (most similar to colicin E9 immunity). Klebicin C has a klebicin D-like receptor-binding domain but a colicin E1-like translocation domain and a colicin E4-like DNase domain (Chavan et al. 2005). Conversely, klebicin D shows a colicin D-like rRNase domain. Similarly, alveicins A and B of *Hafnia alvei* have similar translocation and killing domains but their receptor-binding domains are clearly different (Wertz and Riley 2004).

The klebicin killing domains are homologues of known killing domains – klebicins B, C and D are similar to pyocin S1, colicin E4 and colicin D respectively. These examples exhibit the diversifying selection model proposed for their colicin counterparts, as their evolutionary history seems to involve mutations in their immunity genes in conjunction with corresponding changes in their killing domains, as predicted in the diversifying selection mechanism.

The influence of diversifying recombination is not limited to the closely related bacteriocins of enteric bacteria. The S pyocins of *P. aeruginosa* are speculated to have evolved as a result of recombination between several pore-former and nuclease colicins with other, as yet uncharacterized bacteriocins (Sano et al. 1993a, 1993b). Thus, altering the domain structure of the protein, as seen for pyocins which have switched the receptor recognition and translocation domains relative to the order found in colicins, has not limited the influence of diversifying recombination.

3.5 Evolution of Colicin Killing Domains

The killing domains are the function modules of colicins which catalyze deleterious changes in the sensitive cell, leading to death. Various killing mechanisms are employed by colicins, as described in the sections above. Sequence data of klebicins, S-type pyocins, alveicins, marcescin and cloacins indicate that CLBs exploit the same killing functions as those used by colicins.

Alignments of selected colicin and CLB DNase, rRNase and pore-forming domains are shown in Fig. 3.5. Based on these alignments, Tables 3.1 and 3.2 show percentage similarities and divergence between various pairs of DNase, 16S rRNase and pore-forming colicins respectively. A phylogenetic tree inferred from sequences of nuclease domains of colicins, CLBs and pyocins is

a. DNase killing domains

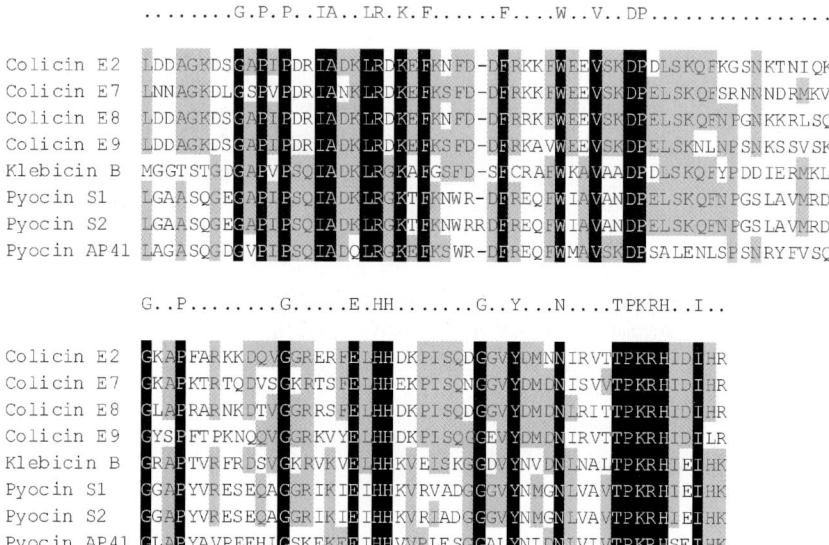

b. DNase immunity proteins

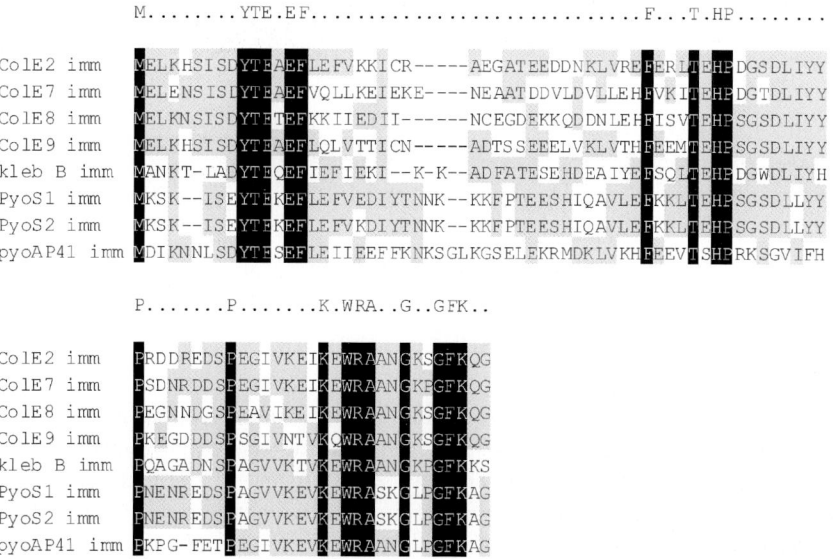

Fig. 3.5 Amino acid alignments of killing domains and immunity proteins of colicins and colicin-like bacteriocins. Killing domains of DNase (**a**), 16S rRNase (**c**), and pore-forming (**e**) domains and their respective immunity proteins (**b**, **d**, **f**) are shown. Amino acid residues conserved in the majority of sequences are highlighted in *grey* whereas those conserved in all sequences are shown in *black*

c. 16S rRNase killing domains

d. 16S rRNase immunity

Fig. 3.5 Continued

shown in Fig. 3.6. The outcome clustered different nuclease activities – DNase domain, rRNase activity, colicin D-like tRNase activity and colicin E5-like tRNase activity – into specific groups. DNase domains of pyocins S1, S2 and AP41 are homologous to killing domains of DNase colicins. Pyocins (S1, S2 and AP41) form a separate cluster to that of colicin DNase domains (E2, E7, E8, E9), with klebicin B clustering closer to pyocins than to colicins (Riley et al. 2001). The corresponding immunity proteins of pyocins S1, S2 and AP41 and colicins E2, E7, E8 and E9 also exhibit significant similarities with each other. Thus, the DNase domain–immunity gene combination of colicins and pyocins appears to share an evolutionary pathway.

e. Pore forming domains

```
                     .................................G........A.........K...D.........
Colicin A     VAEKAKDERELLEKTSELIAGMGDKIGEHLGDKYKAIAKDIADNIKNFQGTIRSFDEAMASLNKITANEAMKINKADEDALVNAWKHVDAQLMANKLGN
Colicin U     AEEKANDEKAVLTKASEIIISVGDKAGEYLGDKYKVLSREIADNIKNFQGTIRSYDEAMASVNKLMANFDLKINAADRDAIVNAWKFAFDAEDMGNKFAA
Colicin B     KKEQENDEKTVLTKTSEVIISVGDKVGEYLGDKYKALSREIAENIDNFQGTIRSYDEQMSSINRKLMANFSLKINATEKEAIVNAWKAFNAEDMGNKFAA
Colicin S4    SMNPDRIQSDVLNKAAEVISDIGNKVGDYLGDAYKSLARFIADDVKNFQGTIRSYDEAMASLNHVLSNFGEFNRADSDALANVWRSIDAQLMANKLGN
Colicin Y     AKAKANDEKAVLTKASEIIISVGDKVGEYLGDKYKALSREIAQNIKNFQGTIRSYDEEQIASVNKLMANFDLRINAKDKEAIVNAWKHVNATITANKLGN
Colicin N     KEEKEKNEKEALLKASELVSGMGDKLGEYLGVKYKNVAKEVANDINFHGRNIRJYNEAMASLNRVLANFPMKVNKSKDAIVNAKWKQVNAKDMANKIGN
Marcescin A   EAARGRDEKDALMKTSELVAAAGEKVGEHLGDKYKAVARQIADDINFQGKRLRSFDEAMASLNKITLNPAMKINKADKEAIVNAWKHVNATITANKLGN
Cloacin 647   ANKKKLTEKELLIKTADLIQDAGEKVSGIASAKYKSLAKELADNVRNFQGNTRSFNDAMKTLNRLTANPMKKISATEKAALINAWKSIDRNNMASRLAN
Colicin Ib    EEKRKRDEINMVKDAIKLTSDFYRTIYDEFGKQASELAKELASVSQ----GIQIKSVDEALNAFDKFRNINKKYMIQDKMAISKALEAINQVHMAENFKL
Colicin Ia    EEKRKQDELKATKDAINFTTEFLKSVSEKYGAKAEQLAREMAGQAK----GKTIRNVEDALKTYEKYRADINKKINAKDKEAIAKAALESVKLSDISSNLNR
Alveicin A    KNAINDNSPNVLQDAIKFTADFYKEVFNAYGEKAEKLAKILLADQAK---GKIRNVEDALKSYEKHKANINKKINAKDKEAIAKALESMDVGKAAKNIAK
Alveicin B    RDAIN-NEKEAVKDAVKFTADFYKEVFKVYGEKAEKLLADQAK-------GKVRNVEDALKSYEKYKTNINKKINAKDKEAIAKALESMDVGKAAKNIAK
Colicin 10    KEAKDALEKSQVKDSVDTMVGFYQYITSQYGEKYSRIAQDLAEKAN----GSKFNSVDEALAAFEKYKNVLDRKFSKVDEDDIFNALESITYDEWAKHLEK
Colicin K     KEAQDALEKSQIKDAVDTMVGFYQYITSQYGEKYAKIAQDLAEKS-----GKIQGVDEALAAFEKYKNVLDKKFSKVDDAIFNALESVNYDELSLNLTK
Colicin 5     KEAKDALEKSQVKDSVDTMVGFYQYITSQYGEKYAKIAQDLAEKS-----GKIQGVDEALAAFEKYKNVLDKKFSKVDDAIFNALESVNYDELSLNLTK
Colicin E1    KKAQNNLLNSQIKDAVDATVSFYQTLTEKYGEKYSKMAQELADKSK----GKIGNVNEALAAFEKYKDVLNRKFSKADEDAIFNAIASVKYDEWAKHLDQ
28b           FVEYQNAELKAIKDGVSLAAGINKDIAEKIGAKYAKLAHDLEAGIQ----GKYIRNYQDAEKTYEQLTKGLNFRLKAQDKAAIVAWLKMIDAEQYARNARV
Cloacin 683   EKNAQDIFSSFPTNAISIASDEHKAVADKFGEKSKEAKALAEASK-----GKIRNAAEFIKAFDKYKNGAINKMYGVAIDQAIANALESLDKKQMAQLSH

                     .................................T..W.......E...................................
Colicin A     LSKAFKVALVVMKVEKVREKSIEGYETGHWGPIMLEIESWVLSGIASSVALGIFSATLGAYALSLGVPAIAVGIAGILLAAVVGALIDDKFADALNNEI
Colicin U     LGKTEKAADYYMKANNVREKSIEGYETGHWGPIMLEIDESWVLSGIASAVALSFFSAIFGTFAMLGVFSTSLAGILAVIAGLVGALIDDNFVDKLNNEI
Colicin B     LGKTFKAADYAIKANNIREKSIEGYETGHWGPIMLEIESWVISGMASAVALSFLSLTLGSALIAFGLSATVVGFVGVVIAGAIGAFIDDKFVDELNHKI
Colicin S4    ISKAFKFADVVMKVEKVREKSIEGYETGHWGPIMLEIESWVLSGIASAVALGVFSATLGAYLSIKLADPAIAVGIVGILLAAVVGALIDDKFADAINNEI
Colicin Y     LGKTFKAADYVMKANNVREKSIEGYQGHWGPIMPEVESWVVSGIASAVALAIFSATLGAYLLAVGASAAVVGIEIGIIIASFIGALIDDKFIDRLNNEI
Colicin N     LGKAFKVADLAIKVEKIREKSIEGYNTAGNWGPILLEVESWIIGGVVAGVAISLFGAVLSFLPIS-GLAVTALGVIGIMTISYLSSFIDANRVSNINNI
Marcescin A   LARAFKVAFGPTGKVIERYDVAVELQKAVHDNWRPFFVKIE-------SLAAGRAASAVTSAWFSVMLGTFVGILGFAIIMAAVSALVNDKFIEQVNELI
Cloacin 647   LSKAFTVAGWLSKVEKVTEKSIIGYATAGMWGPILEVESWVLSGITSALALAVLSLIVSTFLIAGSLPATVTLIAGTLGIYVASLIDDKVAEKVNSQC
Colicin Ib    FSKAFGFTGKVIERYDVAVELQKAVHDNWRPFFVKIE-----------TIIAGNAATALVADVFSILTGSALGIYGYGLLMAVTGALIDESLVEFANKFW
Colicin Ia    FSRGLGYAGKFTSLADWITEFGKAVRHKPWWKPLLFVKTE---------TIIAGNAATALVADVFSILTGSALGIYGYGLLMAVTGALIDESLVEFANKFW
Alveicin A    FSRGLGWVGPAIDITDWFTELYKAVKTDKMRSLYVKTE----------TIAVGLAATHVTALAFSAVLGGFPIGILGYGLIMAGVGALVNETIVDEANKVI
Alveicin B    FSKGLGWVGPAIDITDWFTELYKAVKTDMMRSFYVKTE----------TIAVGLAATHVAALAFSAVCGPVGILGYGLIMAGVGALVNETIVDEANKVT
Colicin 10    ISPAIKVTGYLSFGYDVWDGTLKGLKTGDMRPLFVTIE----------KSAVDFGVAKIVAIMFSFIVGAFPLGFWGIATIGIVSSYIGGDELNKLNELL
Colicin K     ISKSLKITSRVSFLYDVGSDFKNAIEQGMMRPLFVTIE----------KSAVDVGVAKIVAIMFSFIVGVPLGFWGIATVIGIVSSYIGDEINKLNELL
Colicin 5     ISKSLKITSRVSFLYDVGSDFKNAIEQGMMRPLFVTIE----------KSAVDVGVAKIVAIMFSFIVGVPLGFWGIATVTGIVSSYIGDDEINKLNELL
Colicin E1    FAMYLKITGHVSFGYDVVSDILKIKLTGDMRPLFITIE----------KKAADAGVSYVVALLFSLLAGITLGIWGIATVTGICSYIDKNKLNTINEVL
28b           LGRVTGVDWAIKGADLVNAAIEGFSIGMWKAFRNQIEALGL-------SIGAGYTLSAIAAPFAPTLVSSTVGIFAPAYLFGWATSYIDAERAGELEKWV
Cloacin 683   FGRMFAVSETLQWSGFIMGIVKGFRFGDWDAIMG-----TE-------KIAASKLASYMVVAFGAIATTPIGIIGFAAILAITSALITDDLMKKMDLI
```

f. Pore forming immunity

```
                     .........................................................................
Colicin A imm    ---MMNEHSIDTDNRKANNALYLFIIIGLIPLLCIFVVYYKTPDALLLRKIATSTENLPSITSSYNPLMTKVMDIYCKTAPFLAILYILTFKIR
Colicin U imm    ---MQKEH---NDNKQLNNMLSWIAFAGTIPFFIILIICKLNPNLYLTNIIPDSIINIPSVISSYNPIMTKLMDIYGKSAPLLAIITFIIQFRH
Colicin B imm    ---MTSNK---DKNKKANEILYAFSIIGIIPIMAILILRINDPYSQVLYYLYNKVAFLPSITSLHDPVMTTLMSNYNKTAPVMGILVFLCTYKT
Colicin S4 imm   ---MMIDEHSIDIDNRKANNLLYLFMVIGVIPLLCILAVYYTNPDNLFLHTIATSTENIPSITSAYNPLMTKVMDIYCKTAPSLAILFILTFKTR
Colicin Y imm    ---MRKEH---TDNKDLNRTLFLITFAGITPLIIFITYTSNPKPLYLISIIFDNTQNIPSIISAYNPVMTKVMDIYGKSAPLLAIAFTLQLRDR
Colicin N imm    ----------------------------------------MHNTLLEKIIAYLSLPGFHSLNNPPLSEAFNLYVHTAPLAATSLFIFTHKE
Marcescin A imm  MRVNINMYHEIEEGNNIAKKMLCLLLVALVPLLLIAALYYINPKSNFLQGVSEYTTFLPAIVSSNNPLFSKVMDIYLKTSPLSPLFSFFFSFYKRI
Colicin Ib imm   --------------------------------------------------------MKLDISVKYLLKSLIPILILTVFYLGWKDN
Colicin Ia imm   --------------------------------------------------------MNRKYYFNNMWWGWVTGGYMLYMSWDY
Alveicin A imm   --------------------------------------------------------MTEKRTKMLTLNVTPDEHE
Alveicin B imm   --------------------------------------------------------MTRKYYIHNMFWGYFMAVCILYASYGD
Colicin 10 imm   --------------------------------------------------------------------MTVKYYLHNLLE
Colicin K imm    --------------------------------------------------------------------MHKYYLHNLPES
Colicin 5 imm    --------------------------------------------------------------------MHKYYLHNLPES
Colicin E1 imm   --------------------------------------------------------MSLRYYIKNILFGLYCTIYIYLITKNS

                 .........................................................................
Colicin A imm    KLINNTDRNTV--LRSCLLSPLVYAAIVYLFCFRNFELTTAGRPVRLMATNDATLLLFYIGLYSIIFFTTYITLFTPVTAFKLLKKRQ
Colicin U imm    KLEKGINREKL--ITACILTPFIYVFYAYFFLWNNFELTTSGRTVRWMSENDEFTLLFFFICMYYCSFFMTYILCYVPVAVYKIWKER
Colicin B imm    EIIKPVTRKLV--VQSCFWGPVFYAILIYITLFYNLELTTAGGFFKLLSHNVITLFILYCSIYFTVLTMTYAILLMPLLVIKYFKGRQ
Colicin S4 imm   KLIKKTNRNAV--LRSCLLSPFACAVFLYLLCFRNLELTTAGRPARLMTNNDATLLLFYIGLYLIIFFISYFTLFTPVTTFKLLKERQ
Colicin Y imm    KLETIANREKL--ITASIFSPFFYAFYAYFFLWNNFELTTAGRTVRWMSDNDFTLFIFYACLYFCSFFMTYALCYIPVVSYKLWKER
Colicin N imm    ELKPKSSPLRA--LKIILTPFTILYISMYCFLLTDTELTLSSKTFVLIVKKRSVFVFFLYNTIYWDIYIHIFVLLVPYRNI
Marcescin A imm  KLKSNQSVSKL--LVTFIFFTXFYVCLTGFLFTNIELTNSVRTLKAMSTNDITLFLFYITLYAGIYVFGCLYLWFGIGTVKAFKARQRTTSL
Colicin Ib imm   QENARMFYAFIGCIISAITFFFSMRIIQKMVIRFTGKEFWQKDFFTNPVGGSLTAIFELFCFVISVPVVAIYLIFILCKALSGK
Colicin Ia imm   FKYRLLFWCIS--LCGMVLYPVAKWYIEDTALKFTRPDFWNSGFFADTPGKMGILAVYTGTVFILSLPLSMIYILSVIIKRLSVR
Alveicin A imm   LLARCDSPRLATWMREVCLEEKPARTSKPLPTLAPEILPQLAGMGNLNQIARRLNSGEWSAHDRVQVVAALLTLERELQFLREQAR
Alveicin B imm   EPKIIALRIFG--LASAILFFFSRFLIEKTALRYTKKEFWETGFFKDGVPKTYLMTLYLIFIFMTSIPIGVISVFFEIKNVTAKL
Colicin 10 imm   LIPWLFYLLLN--YKTPPFSLIIFIASIHVLYPYSKLTIF-SFIQNTTMKMKKEPWYSYNLFYFLYLAMAIPVGLPSFIYYSLKRN
Colicin K imm    LIPWILILIFN--DNDNTPLLFIFISSIHVILYPYSKLTIS-RYIKENTKLKKEPWYLCKLSALFYLLMAIPVGLPSFIYYTLKRN
Colicin 5 imm    LIPWILILIFN--DNDNTPLLFIFISSIHVILYPYSKLTIS-RYIKENTKLKKEPWYLCKLSALFYLLMAIPVGLPSFIYYTLKRN
Colicin E1 imm   EGYYFLVSDKM--LYAIVISTILCPYSKYAIEYIAFNFIKKDFFERRKNLNNAPVAKLNLFMLYNLLCLVLAIPFGLLGLFISIKNN
```

Fig. 3.5 Continued

Table 3.1 Pairwise distance of DNase killing domains and associated immunity proteins from colicins and colicin-like bacteriocins

	Colicin E2, colicin E2 imm	Colicin E7, colicin E7 imm	Colicin E8, colicin E8 imm	Colicin E9, colicin E9 imm	Klebicin B, klebicin B imm	Pyocin S1, pyocin S1 imm	Pyocin S2, pyocin S2 imm	Pyocin AP41, pyocin AP41 imm
DNase killing domains (percent similarity)								
Colicin E2	***	71.8	82.7	77.3	47.3	50.0	50.9	43.6
Colicin E7	35.3	***	71.8	65.5	50	47.3	48.2	42.7
Colicin E8	19.7	35.3	***	74.5	50.9	53.6	54.5	48.2
Colicin E9	27.1	46.1	31.1	***	44.5	46.4	47.3	45.5
Klebicin B	87.4	79.9	77.5	95.7	***	58.2	59.1	46.4
Pyocin S1	79.9	87.4	70.6	90.1	60.4	***	99.1	57.3
Pyocin S2	77.5	84.8	68.5	87.4	58.4	0.9	***	57.3
Pyocin AP41	98.7	101.7	84.8	92.9	90.1	62.3	62.3	***
Associated immunity proteins (percent divergence)								
Colicin E2 imm	***	67.4	57.6	68.6	56.6	60.7	61.9	49.4
Colicin E7 imm	42.6	***	62.4	59.3	50	49.4	50.6	46.5
Colicin E8 imm	61.5	51.9	***	55.3	49.4	50.6	49.4	48.8
Colicin E9 imm	40.6	58	66.8	***	51.8	52.4	52.4	49.4
Klebicin B imm	63.7	79.9	81.5	75.2	***	56.0	54.8	46.4
Pyocin S1 imm	55.1	81.4	78.3	73.7	65.3	***	98.9	46.5
Pyocin S2 imm	52.8	78.3	81.5	73.7	68.0	1.2	***	45.3
Pyocin AP41 imm	81.4	89.7	83.1	81.4	89.9	89.7	93.2	***

Table 3.2 Pairwise distance of 16S rRNase killing domains and associated immunity proteins from colicins and colicin-like bacteriocins

	Cloacin DF13, cloacin DF13 imm	Colicin E3, colicin E3 imm	Colicin E4, colicin E4 imm	Colicin E6, colicin E6 imm	Klebicin C, klebicin C imm	Klebicin CCL, klebicin CCL imm
16S rRNase killing domains (percent similarity)						
Cloacin DF13	***	84.2	85.3	91.6	84.2	87.4
Colicin E3	17.8	***	83.2	89.5	83.2	87.4
Colicin E4	16.5	19.1	***	84.2	90.5	90.6
Colicin E6	9.0	11.4	17.8	***	85.3	85.3
Klebicin C	17.8	19.1	10.2	16.5	***	85.3
Klebicin CCL	13.9	13.9	10.0	16.5	16.5	***
Associated immunity proteins (percent divergence)						
Cloacin DF13 imm	***	70.6	76.5	78.8	66.7	78.8
Colicin E3 imm	37.3	***	74.1	87.1	63.1	80
Colicin E4 imm	28.3	31.8	***	71.8	77.4	83.5
Colicin E6 imm	24.9	14.2	35.4	***	63.1	78.8
Klebicin C imm	43.9	50.5	27.0	50.5	***	66.7
Klebicin CCL imm	24.9	23.3	18.6	24.9	43.9	***

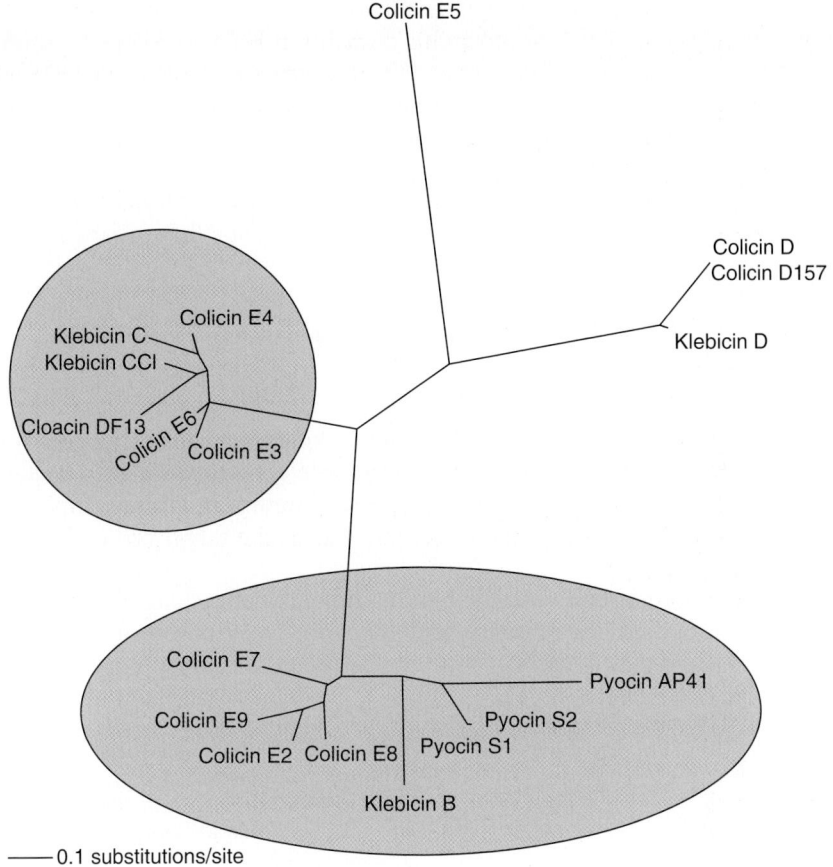

Fig. 3.6 Maximum-likelihood phylogram of nuclease killing domains of colicins and colicin-like bacteriocins. Bootstrap values >50% are indicated

The rRNase domains (colicins E3, E4 and E6, cloacin DF13, klebicin D and CCL) form a tight cluster with very short branches in the inferred phylogenetic tree (Fig. 3.6). This domain is highly conserved (Tables 3.1 and 3.2, Fig. 3.5), compared to the DNase and pore-forming domains. The tRNase domain of colicin D and klebicin D are closely related to each other whereas that of colicin E5, which has a t-RNase activity distinct from that of the former two, forms a stand-alone branch.

Pyocin S3 has DNase activity but has no sequence homology to DNase domains of colicins or other pyocins currently characterized (Duport et al. 1995). Therefore, it can be speculated that the pyocin S3 killing domain has evolved independently from that of other DNase domains. Its cognate immunity protein, which is larger than other DNase immunity proteins, appears to have co-evolved with the new DNase domain to provide protection to the pyocin S3-producing cell.

A maximum-likelihood tree inferred for the pore-forming domains is shown in Fig. 3.7. The pore-forming domains form two distinct groups: type A (e.g., colicins E1, 5, 10, K, Ia and Ib, alveicins A and B, cloacin 683) and type B (colicins A, B, N, S4, U and Y, marcescin A and cloacin 647). A bacteriocin from *Serratia*, marcescin A, is most closely related to colicins N and A in terms of level of similarity. However, colicin A clusters with colicins Y, U and S4, with which it has a higher level of similarity. Alveicins A and B produced by *Hafnia alvei* are related to colicin Ia, a type B pore former.

3.6 Evolution of the Translocation and Receptor-Binding Domains

Nature has opted for multiple mechanisms to kill target cells using pore formation, DNase, rRNase, tRNase or muramidase activity. Similarly, bacteriocins have exploited multiple species-specific receptors and import functions found in bacterial cells to mediate recognition of the target cell and access target molecules.

The translocation and receptor-binding domains of colicins do not show significant cross-species presence and function, as is observed for killing

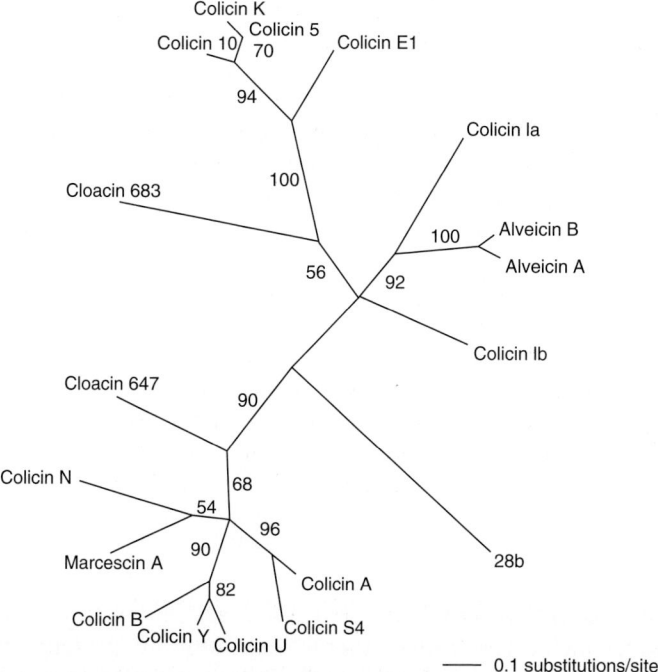

Fig. 3.7 Maximum-likelihood phylogram of pore-forming killing domains of colicins and colicin-like bacteriocins. Bootstrap values >50% are indicated

domains which are freely exchanged between colicins and CLBs. Thus, killing domains appear to be transferred more frequently than the target-defining (specifying) domains for translocation and receptor binding. The CLBs have no (or significantly lower) similarity with well-characterized translocation and receptor-binding domains of colicins, in contrast to that observed for the killing domains (Riley et al. 2001; Wertz and Riley 2004; Chavan et al. 2005). These domains have evolved in CLBs so as to be more adept in exploiting the receptors and translocation mechanisms of their own species. This has resulted in the evolution of colicins and CLBs which kill very closely related bacteria. It appears that within their ecological niches, bacteria selectively develop toxins to control closely related populations which have similar environmental requirements, rather than addressing competition from different genera. In other words, bacterial strains appear to be more threatened by siblings than by cousins.

There are some exceptions to the narrow killing ranges seen in most CLBs. Killing activity across a genera does exist, one such example being 28B produced by *Serratia marcescens* which targets *E. coli*. The sequence of the translocation domain of 28B is remarkably similar to that of colicins A and U, which also kill *E. coli*.

The translocation domains of S-type pyocins, colicins E2, E3 and cloacin DF13 are similar to each other. However, chimeric constructions of pyocin and colicin domains do not result in expected exploitation of the translocation system across the species. Thus, these domains are rather species-specific, and their similarities to each other appear to have no functional significance (Kageyama et al. 1996).

Marcescin A, also produced by *Serratia marcescens*, appears to have a novel translocation domain which has probably diversified to utilize a parallel translocation system or employ a different system in sensitive strains of *Serratia* (unpublished data). Such species specificity of translocation domains was also observed in alveicins A and B (Wertz and Riley 2004) and klebicins (Riley et al. 2001; Chavan et al. 2005). Similar species-specific diversity is found in their receptor-binding domains. Whereas the associated receptors have been characterized for all colicins, those targeted by CLBs have not been investigated to date. The receptor-binding domains of klebicins C and D have 45% protein sequence similarity to the corresponding domain of colicin D, indicating that these bacteriocins bind to similar proteins in their target species. On the other side of the spectra, the receptor-binding domains of klebicin B, alveicins A and B, and marcescin A are considered to be novel.

3.7 Evolution of Colicin Regulatory Sequences

Colicins are expressed from a SOS response regulated promoter. SOS regulation of a gene is mediated by binding of LexA repressor to a 20-bp palindrome (5′-TACTGTATATATATACAGTA-3′) sequence, referred to as the LexA-binding box (Walker 1984). However, regulatory regions of colicins contain two overlapping LexA-binding boxes, which is considered a characteristic feature of a colicin promoter (Parker 1986; Riley et al. 2001). The only instances of non-colicin genes with colicin-like dual-overlapping LexA-binding boxes are found in three prophage genes in *S. typhi* and *S typhimurium* chromosomes. There are a few exceptions among colicins – cloacin DF13, marcescin A and klebicin D all have single LexA-binding boxes, and colicins Ia and Ib contain a second degenerate LexA box (Varley and Boulnois 1984). LexA repressor binding regions are absent in regulatory sequences of pyocins and bacteriocin 28B, even though these toxins are inducible by DNA-damaging agents. Pyocins are regulated by two genes, *prtN* and *prtR* (Matsui et al. 1993). In the case of bacteriocin 28B, a CLB produced by *Serratia*, SOS induction is mediated indirectly by a transcriptional activator protein of which the expression is repressed by SOS regulation (Ferrer et al. 1996).

A comparison of promoter sequences starting from the –35 box through to the start codon of colicins and CLBs reveals few highly conserved regions. The –35 box is most conserved and matches the consensus (5′ TTGACA 3′) whereas the –10 box shows only 50% match to the consensus (5′ TATAAT 3′; Fig. 3.8). The LexA-binding box is another conserved element which is followed by T-rich segments, and finally by the RBS (ribosome binding site). Most of the colicin promoters cluster together whereas the CLB promoters are quite divergent. The promoters of colicin A, klebicin B and a newly sequenced cloacin 683 (unpublished data) have a leader sequence between the LexA-binding boxes and the RBS. Interestingly, all these bacteriocins were sequenced from non-*E. coli* species. Klebicin B and cloacin 683 are expressed by *Klebsiella pneumoniae* and *Enterobacter cloacae* respectively, and colicin A was isolated from *Citrobacter freundii*.

Among all CLBs, the pesticin (produced by *Yersinia pestis*) promoter is most closely related to colicin promoters. The alveicin A/B promoters also share significant similarity with colicins. The bacteriocin 28B promoter is very different from the colicin promoters, though its translocation domain is similar to that seen in colicins A and U.

3.8 Colicin D: A Possible Intermediate Between Pyocins and Colicins

The modular design of colicins and pyocins provides flexibility for generating new bacteriocins by recombination mechanisms. As described above, colicin B is a classic example, with colicin D-like translocation and receptor-binding domains and a colicin A/Y-like pore-forming domain. Similarly, translocation and DNase domains of pyocins S1 and S2 are homologous whereas their receptor-binding domains have been acquired from different sources. Thus,

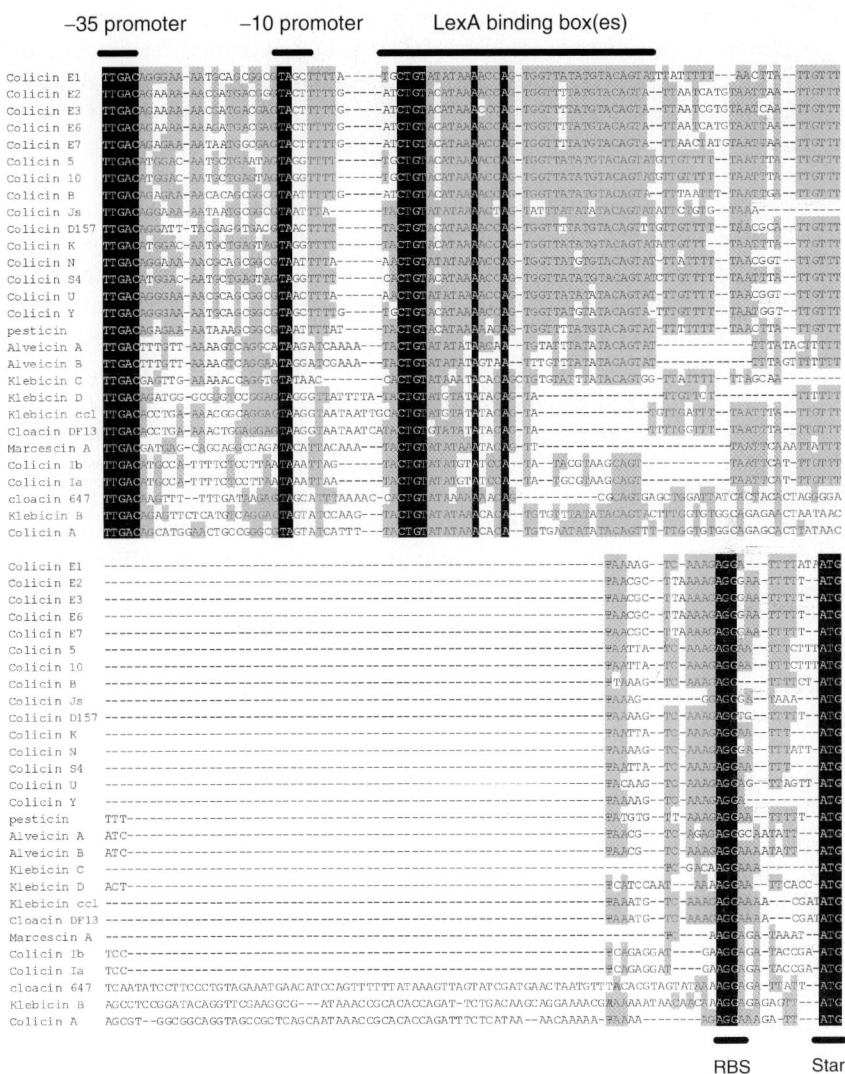

Fig. 3.8 Nucleotide alignment of regulatory regions of colicins and colicin-like bacteriocins. Nucleotides conserved in all or a majority of sequences are highlighted in *black* or *grey* respectively. The positions of −35 and −10 promoter regions, the LexA-binding regulatory sequence, the RBS (ribosomal binding Shine Dalgarno region) and start codons are indicated

nature has created a pool of candidates for each domain which can be mixed with one another. The cytotoxic domains are obviously shared between colicins and pyocins. However, sequences of colicin D and other colicin-like bacteriocins indicate that there has been more sharing between the different bacteriocins. The N-terminal 313 amino acid residues of colicin D are 95%

identical to the colicin B translocation and receptor-binding domains. The tRNase domain of colicin D consists of 97 aa residues in its C-terminal region (Takahashi et al. 2006). Thus, colicin D has a complete set of domains needed for its function. However, it contains an additional domain (~290 aa residues) between the central receptor-binding domain and the cytotoxic domain which has some similarity with the translocation domain of pyocins S1, S2, S3 and AP41. In all cases (colicin D and pyocins), this domain is located towards the N-terminus of the killing domain and hence, it is tempting to hypothesize that a translocation domain:cytotoxic domain combination was transferred from one bacteriocin to another (most probably from pyocin to colicins or CLBs). A similar organization is observed in klebicins B, C and D, all of which have a pyocin translocation-like domain adjacent to their killing domain. Klebicin B also has significant levels of similarity with the N-terminal translocation domain of colicin A whereas klebicin D has a colicin E1-like translocation domain. These klebicins also have a regulatory region similar to that of colicins. It is not clear if the original function of this second domain has been retained or lost. Another interesting feature of klebicin bacteriocins observed recently is the presence of an additional gene in the klebicin operon. This gene is located upstream of the activity gene in both the klebicin C and D operons. The encoded protein shares sequence similarity to the C-terminal domain of the phage tail fibre-like gene. Experimental evidence indicates that it is required for klebicin D activity (Chavan et al. 2005). This reinforces the idea that nature is continuing to experiment with different killing strategies.

The function of a second domain found in pyocins S2, S3 and AP41 also has not yet been determined. It could be a vestigial domain, similarly to the one described above for colicin D and klebicins. Perhaps, as a phage evolves to specialize as an R- or F-type pyocin, these new domains/genes provide additional function to increase the fitness of the host cell, thereby giving them an edge over their competitors.

3.9 Conclusions

Colicin expression results in lysis of the producer cell, due to co-expression of a lysis protein which mediates the release of colicin into the extra-cellular environment. It is obvious that such a suicidal mechanism is costly for the cell, and therefore raises questions regarding the need for bacterial cells to maintain colicin-encoding genes. One hypothesis is that colicins are being used by selfish plasmid systems for their own maintenance, as a parasite in bacterial hosts. Those cells which cure themselves of the plasmid also lose immunity and thus are killed by the colicins produced by neighbour siblings. On the other hand, colicins have also been implicated as a defence mechanism in competition for niche space and resources.

In some cases, the colicin and lysis genes form an operon tightly regulated by the SOS system, which responds to significant DNA damage. Thus, induction of colicin and co-induction of lysis genes may occur only in damaged cells, resulting in the cell's death, and thus may be considered similar to apoptosis in eukaryotic cells. Experimental evidence suggests that expression of colicin is induced in only a small fraction of the population (Mulec et al. 2003). These colicin-expressing cells eventually die but produce enough colicin to kill related, but competing, cells. Thus, a fraction of colicin-harbouring cells display altruistic behaviour by "sacrificing themselves" for the larger benefit of their clonal kin. The altruistic behaviour of colicin producers has elicited considerable interest and warrants further consideration.

The narrow killing spectra of colicins and CLBs provide additional points of consideration concerning the necessity of these bacteriocins for the cell. If the colicin-harbouring plasmid was indeed a selfish parasitic system, then these bacteriocin genes should have moved to multiple genera by a series of horizontal transfer events, as seen with antibiotic-resistant genes. For example, the plasmid encoding klebicin B could have moved from *Klebsiella pneumoniae* to *Pseudomonas* sp. or *Citrobacter* sp. However, killing activities across the genera are uncommon. Rather, we observe kin discrimination by bacteriocin-producing strains. So, why should colicin-producing *E. coli* kill its own kin, rather than individuals from a different species or genus? Maximum competition is observed between cells with more similar nutritional and niche requirements. In addition, the presence of a diverse assemblage of other bacterial species and genera is critical for the health of the environmental niche as a whole. Since colicins, and numerous other bacteriocins, have persisted for millions of years, the benefit of these actions must outweigh the consequences of cells being sacrificed in the process.

References

Baquero F, Bouanchaud D, Martinez-Perez MC, Fernandez C (1978) Microcin plasmids: a group of extrachromosomal elements coding for low-molecular-weight antibiotics in *Escherichia coli*. J Bacteriol 135:342–347

Berdy J (1974) Recent developments of antibiotic research and classification of antibiotics according to chemical structure. Adv Appl Microbiol 18:309–406

Bowman CM, Dahlberg JE, Ikemura T, Konisky J, Nomura M (1971) Specific inactivation of 16S ribosomal RNA induced by colicin E3 in vivo. Proc Natl Acad Sci USA 68:964–968

Braun V, Pilsl H, Gross P (1994) Colicins: structures, modes of action, transfer through membranes, and evolution. Arch Microbiol 161:199–206

Chak KF, Kuo WS, Lu FM, James R (1991) Cloning and characterization of the ColE7 plasmid. J Gen Microbiol 137:91–100

Chavan M, Rafi H, Wertz J, Goldstone C, Riley MA (2005) Phage associated bacteriocins reveal a novel mechanism for bacteriocin diversification in *Klebsiella*. J Mol Evol 60:546–556

Duport C, Baysse C, Michel-Briand Y (1995) Molecular characterization of pyocin S3, a novel S-type pyocin from *Pseudomonas aeruginosa*. J Biol Chem 270:8920–8927

Ferrer S, Viejo MB, Guasch JF, Enfedaque J, Regue M (1996) Genetic evidence for an activator required for induction of colicin-like bacteriocin 28b production in *Serratia marcescens* by DNA-damaging agents. J Bacteriol 178:951–960

Fredericq P (1957) Colicins. Annu Rev Microbiol 11:7–22

Fredericq P (1963) On the nature of colicinogenic factors: a review. J Theor Biol 4:159–165

Gordon DM, Riley MA (1999) A theoretical and empirical investigation of the invasion dynamics of colicinogeny. Microbiology 145(3):655–661

Guasch JF, Enfedaque J, Ferrer S, Gargallo D, Regue M (1995) Bacteriocin 28b, a chromosomally encoded bacteriocin produced by most *Serratia marcescens* biotypes. Res Microbiol 146:477–483

Hopwood DA, Chater KF (1980) Fresh approaches to antibiotic production. Philos Trans R Soc Lond B Biol Sci 290:313–328

Kageyama M, Kobayashi M, Sano Y, Masaki H (1996) Construction and characterization of pyocin-colicin chimeric proteins. J Bacteriol 178:103–110

Kim SR, Maenhaut-Michel G, Yamada M, Yamamoto Y, Matsui K, Sofuni T, Nohmi T, Ohmori H (1997) Multiple pathways for SOS-induced mutagenesis in *Escherichia coli*: an overexpression of dinB/dinP results in strongly enhancing mutagenesis in the absence of any exogenous treatment to damage DNA. Proc Natl Acad Sci USA 94:13792–13797

Kirkup BC, Riley MA (2004) Antibiotic-mediated antagonism leads to a bacterial game of rock-paper-scissors in vivo. Nature 428:412–414

Kuroda K, Kagiyama R (1983) Biochemical relationship among three F-type pyocins, pyocin F1, F2, and F3, and phage KF1. J Biochem (Tokyo) 94:1429–1441

Masaki H, Ogawa T (2002) The modes of action of colicins E5 and D, and related cytotoxic tRNases. Biochimie 84:433–438

Matsui H, Sano Y, Ishihara H, Shinomiya T (1993) Regulation of pyocin genes in *Pseudomonas aeruginosa* by positive (prtN) and negative (prtR) regulatory genes. J Bacteriol 175:1257–1263

Michel-Briand Y, Baysse C (2002) The pyocins of *Pseudomonas aeruginosa*. Biochimie 84:499–510

Moreno F, Gonzalez-Pastor JE, Baquero MR, Bravo D (2002) The regulation of microcin B, C and J operons. Biochimie 84:521–529

Mulec J, Podlesek Z, Mrak P, Kopitar A, Ihan A, Zgur-Bertok D (2003) A cka-gfp transcriptional fusion reveals that the colicin K activity gene is induced in only 3 percent of the population. J Bacteriol 185:654–659

Nakayama K, Takashima K, Ishihara H, Shinomiya T, Kageyama M, Kanaya S, Ohnishi M, Murata T, Mori H, Hayashi T (2000) The R-type pyocin of *Pseudomonas aeruginosa* is related to P2 phage, and the F-type is related to lambda phage. Mol Microbiol 38:213–231

Ogawa T, Tomita K, Ueda T, Watanabe K, Uozumi T, Masaki H (1999) A cytotoxic ribonuclease targeting specific transfer RNA anticodons. Science 283:2097–2100

Ohno-Iwashita Y, Imahori K (1982) Assignment of the functional loci in the colicin E1 molecule by characterization of its proteolytic fragments. J Biol Chem 257:6446–6451

Parker RC (1986) Mitomycin C-induced bidirectional transcription from the colicin E1 promoter region in plasmid ColE1. Biochim Biophys Acta 868:39–44

Parret A, De Mot R (2000) Novel bacteriocins with predicted tRNase and pore-forming activities in *Pseudomonas aeruginosa* PAO1. Mol Microbiol 35:472–473

Pilsl H, Braun V (1995) Strong function-related homology between the pore-forming colicins K and 5. J Bacteriol 177:6973–6977

Pugsley AP (1984) The ins and outs of colicins. Part I. Production, and translocation across membranes. Microbiol Sci 1:168–175

Reeves P (1965) The bacteriocins. Bacteriol Rev 29:25–45

Riley MA (1993a) Molecular mechanisms of colicin evolution. Mol Biol Evol 10:1380–1395

Riley MA (1993b) Positive selection for colicin diversity in bacteria. Mol Biol Evol 10:1048–1059

Riley MA (1998) Molecular mechanisms of bacteriocin evolution. Annu Rev Genet 32:255–278

Riley MA, Gordon DM (1992) A survey of Col plasmids in natural isolates of *Escherichia coli* and an investigation into the stability of Col-plasmid lineages. J Gen Microbiol 138(7):1345–1352

Riley MA, Wertz JE (2002a) Bacteriocin diversity: ecological and evolutionary perspectives. Biochimie 84:357–364

Riley MA, Wertz JE (2002b) Bacteriocins: evolution, ecology, and application. Annu Rev Microbiol 56:117–137

Riley MA, Tan Y, Wang J (1994) Nucleotide polymorphism in colicin E1 and Ia plasmids from natural isolates of *Escherichia coli*. Proc Natl Acad Sci USA 91:11276–11280

Riley MA, Cadavid L, Collett MS, Neely MN, Adams MD, Phillips CM, Neel JV, Friedman D (2000) The newly characterized colicin Y provides evidence of positive selection in pore-former colicin diversification. Microbiology 146(7):1671–1677

Riley MA, Pinou T, Wertz JE, Tan Y, Valletta CM (2001) Molecular characterization of the klebicin B plasmid of *Klebsiella pneumoniae*. Plasmid 45:209–221

Riley MA, Goldstone CM, Wertz JE, Gordon D (2003) A phylogenetic approach to assessing the targets of microbial warfare. J Evol Biol 16:690–697

Sano Y, Kobayashi M, Kageyama M (1993a) Functional domains of S-type pyocins deduced from chimeric molecules. J Bacteriol 175:6179–6185

Sano Y, Matsui H, Kobayashi M, Kageyama M (1993b) Molecular structures and functions of pyocins S1 and S2 in *Pseudomonas aeruginosa*. J Bacteriol 175:2907–2916

Schaller K, Holtje JV, Braun V (1982) Colicin M is an inhibitor of murein biosynthesis. J Bacteriol 152:994–1000

Smarda J, Obdrzalek V (2001) Incidence of colicinogenic strains among human *Escherichia coli*. J Basic Microbiol 41:367–374

Stachura I, McKinley FW, Leidy G, Alexander HE (1969) Incomplete bacteriophage-like particles in ultraviolet-irradiated haemophilus. J Bacteriol 98:818–820

Strauch E, Kaspar H, Schaudinn C, Damasko C, Konietzny A, Dersch P, Skurnik M, Appel B (2003) Analysis of enterocoliticin, a phage tail-like bacteriocin. Adv Exp Med Biol 529:249–251

Takahashi K, Ogawa T, Hidaka M, Ohsawa K, Masaki H, Yajima S (2006) Purification, crystallization and preliminary X-ray analysis of the catalytic domain of the *Escherichia coli* tRNase colicin D. Acta Crystallogr Sect F Struct Biol Crystallogr Commun 62:29–31

Tan Y, Riley MA (1996) Rapid invasion by colicinogenic *Escherichia coli* with novel immunity functions. Microbiology 142(8):2175–2180

Tan Y, Riley MA (1997a) Nucleotide polymorphism in colicin E2 gene clusters: evidence for nonneutral evolution. Mol Biol Evol 14:666–673

Tan Y, Riley MA (1997b) Positive selection and recombination: major molecular mechanisms in colicin diversification. Trends Ecol Evol 12:348–351

Toba M, Masaki H, Ohta T (1988) Colicin E8, a DNase which indicates an evolutionary relationship between colicins E2 and E3. J Bacteriol 170:3237–3242

Tomita K, Ogawa T, Uozumi T, Watanabe K, Masaki H (2000) A cytotoxic ribonuclease which specifically cleaves four isoaccepting arginine tRNAs at their anticodon loops. Proc Natl Acad Sci USA 97:8278–8283

Varley JM, Boulnois GJ (1984) Analysis of a cloned colicin Ib gene: complete nucleotide sequence and implications for regulation of expression. Nucl Acids Res 12:6727–6739

Vollmer W, Pilsl H, Hantke K, Holtje JV, Braun V (1997) Pesticin displays muramidase activity. J Bacteriol 179:1580–1583

Walker GC (1984) Mutagenesis and inducible responses to deoxyribonucleic acid damage in *Escherichia coli*. Microbiol Rev 48:60–93

Wertz JE, Riley MA (2004) Chimeric nature of two plasmids of *Hafnia alvei* encoding the bacteriocins alveicins A and B. J Bacteriol 186:1598–1605

4 The Diversity of Bacteriocins in Gram-Positive Bacteria

NICHOLAS C.K. HENG[1], PHILIP A. WESCOMBE[2], JEREMY P. BURTON[2], RALPH W. JACK[1] AND JOHN R. TAGG[1]

Summary

Gram-positive bacteria, and especially the lactic acid bacteria (LAB), are now increasingly studied for their production of bacteriocin-like inhibitory activity. This has yielded detailed insight into many unique features of this surprisingly heterogeneous array of antibiotic molecules, a group apparently united only by their proteinaceous composition and targeted killing of bacteria generally closely related to the producer bacterium. Contemporary developments in this field have included increased knowledge of factors influencing bacteriocin expression, mode of action and specific host-cell immunity. Much of the burgeoning interest in the bacteriocin-producing LAB is driven by their perceived potential practical applications either to food preservation or as probiotics. In this chapter, we propose that all of the currently confirmed bacteriocins of Gram-positive bacteria can be classified into four broad groups: (1) lantibiotics, (2) small non-modified peptides, (3) large proteins, and (4) cyclic peptides.

4.1 Introduction

4.1.1 Bacteriocins: A Historical Perspective

The term antibiotic is generically used to describe substances produced by organisms that selectively interfere with the growth of other organisms. Within this extremely broad category of bioactive molecules, the subset known as the bacteriocins comprises the ribosomally synthesized proteinaceous compounds released extracellularly by bacteria that can be shown to interfere with the growth of other bacteria, typically including some that are

---

[1]Department of Microbiology and Immunology, Otago School of Medical Sciences, University of Otago, P.O. Box 56, Dunedin 9054, New Zealand, e-mail: john.tagg@stonebow.otago.ac.nz; nicholas.heng@stonebow.otago.ac.nz; ralph.jack@stonebow.otago.ac.nz
[2]BLIS Technologies Ltd., Dunedin 9054, New Zealand, e-mail: philip.wescombe@blis.co.nz; jeremy.burton@blis.co.nz

closely related to the producing bacterium and to which the producer cell expresses a degree of specific immunity.

The study of inter-bacterial inhibition, similarly to so many other fundamental facets of microbiology, can trace its origins to Louis Pasteur. In 1877 Pasteur, together with his assistant Joubert, in seeking a way to control the growth of the anthrax bacillus, reported both in vivo and in vitro inhibitory activity associated with co-inoculated "common bacteria" (probably *Escherichia coli*) isolated from urine. Pasteur's pioneering studies heralded several decades of investigations, predating the antibiotic era, which focused upon the dosing of patients with relatively harmless bacteria in an attempt to counter the proliferation of pathogens – the so-called bacterial interference strategy – an approach to infection control now experiencing a renaissance after the half century of neglect that followed the discovery of penicillin, and the associated smug dependence of clinicians on the profligate use of therapeutic non-ribosomally synthesized antibiotics to control bacterial infection. Most of the early successes in defining the nature of bacteriocins related to those of Gram-negative bacteria, especially the colicins, and much of this knowledge stemmed from the work of Gratia and Fredericq. It was Gratia who first described antagonism between strains of *E. coli* (Gratia 1925). Interestingly, the first documented inhibitory strain produces colicin V, a bacteriocin of the microcin class that, in many respects, more closely resembles bacteriocins typically produced by Gram-positive bacteria (Håvarstein et al. 1994). Fredericq used specific (receptor-deficient) colicin-resistant mutants to classify the colicins (Fredericq 1946). General characteristics of the colicins included (1) plasmid-encoded, large domain-structured proteins, (2) bacteriocidal activity via specific receptors, and (3) lethal SOS-inducible biosynthesis. The study of bacteriocins of Gram-positive bacteria got off to a relatively faltering start, largely focusing on the staphylococci, and with various attempts to apply similar principles of classification to those that had been established for the colicins. However, relatively few of the protein antibiotics of Gram-positive bacteria fit closely the classical colicin mold. Major differences include their relatively broad activity spectra, less defined specific producer cell self-protection (immunity), and absence of SOS-inducibility. In the past three decades, studies of bacteriocins of Gram-positive bacteria, especially those of the lactic acid bacteria (LAB), have come to dominate the bacteriocin-related literature, a change largely driven by commercial imperatives.

4.1.2 Bacteriocins of Gram-Positive Bacteria

A landmark observation in the investigation of bacteriocins of Gram-positive bacteria was the documentation in 1947 that some of the inhibitory activity of lactococci (group N streptococci) toward other LAB is due to a molecule characterized as a proteinaceous antimicrobial called "group N inhibitory substance", or nisin (Mattick and Hirsch 1947). Now approved for use as a food additive in around 50 countries, nisin has led the way in the bacteriocin field, not only with regard to the accumulated knowledge of its chemical characteristics and genetic basis but also for the extent and variety of its practical applications. The spectacular commercial success of nisin stimulated a gold rush-like frenzy of prospecting activity for comparable inhibitory agents. In 1976, the first review specifically of the literature on bacteriocins of Gram-positive bacteria noted the beginnings of this groundswell of exploration for nisin-like molecules produced by Gram-positive bacteria, and predicted growing interest in their potential applications to bacterial interference and food preservation (Tagg et al. 1976). On the other hand, failure to find practical outcomes for the bacteriocins of Gram-negative bacteria had caused them to be rather disregarded by researchers and funding organizations.

Two decades later, the scientific literature in the bacteriocin field had become dominated by studies of the bacteriocins of LAB, although the vast majority of these studies did not progress beyond superficial descriptions of their activity spectra against random collections of indicator bacteria and lavish predictions of their potential practical applications (Jack et al. 1995). Some small groups of enthusiasts had continued to explore the use of bacterial interference throughout the early days of the antibiotic era, and most of these studies targeted *Staphylococcus aureus*, due to its predilection for antibiotic resistance development. Some success was achieved with the use of the relatively avirulent 502A strain of *St. aureus*, although the active inhibitory principle(s) was never characterized (Bibel et al. 1983). More recently, the concept of specific modulation of the oral microflora via the introduction of well-characterized bacteriocin producers has found application to the control of a variety of ailments and infections of the oral cavity, ranging from streptococcal pharyngitis and dental caries to otitis media and halitosis (Tagg and Dierksen 2003). Meanwhile, an impressive number of bacteriocin-producing, natural food-associated isolates, mostly LAB and many of GRAS (generally regarded as safe) status, continue to be touted for their potential ability to specifically influence the bacterial content (both beneficial and detrimental) of food – providing a so-called "rudimentary innate immunity to foodstuffs" (Cotter et al. 2005b). Nevertheless, there have still been no comparable commercial successes to that of the benchmark bacteriocin, nisin.

4.1.3 Why Produce Bacteriocins?

Although there has been considerable discussion about their role in nature, it seems that the overriding "raison d'être" for bacteriocins is to provide the producing organism with an ecological advantage over its most likely competitors (Riley and Wertz 2002). So, how common is bacteriocin production?

It has been speculated that all members of the Eubacteria and also of the Archaea, when freshly isolated from their natural ecosystems, are probably equipped with the capability of expressing bacteriocins. It has even been suggested that, where no bacteriocins have yet been found to be produced by a bacterial isolate, it is only because the researchers have not yet defined the expression or detection conditions appropriate to display that strain's bacteriocinogenicity in vitro (Tagg 1992). As many of these bacteriocins are structurally complex, and therefore undoubtedly come at a considerable genetic and biosynthetic cost to the producer cell, they must clearly be functionally indispensable for the existence and persistence in nature of bacterial and archaeal cell lineages. However, under laboratory conditions where bacteria are typically grown as monocultures (with no competitors) and under pampered (low stress) conditions of nutritional excess, the production of anti-competitor molecules (bacteriocins) can more readily be dispensed with. Indeed, exposure of laboratory cultures to chemical curing agents or growth at elevated temperature can lead to the elimination of bacteriocin-encoding plasmids. In other cases, insertion of transposons within bacteriocin loci can lead to loss of bacteriocin expression. Close linkage of bacteriocin structural and immunity determinants on plasmids encourages retention of these plasmids under conditions favorable for bacteriocin expression, since plasmid-cured (and thus bacteriocin-sensitive) derivatives are likely to be counter-selected either in natural ecosystems or in laboratory cultures containing the homologous bacteriocin. Bacteriocin production, however, must inherently be an unstable strain characteristic, otherwise all bacteria would be expected to express multiple bacteriocins – so, from the bacterium's perspective, there must be a tradeoff between the metabolic (and genetic) cost to the cell of bacteriocinogenicity and the survival benefits accrued from the expression of bacteriocin(s) and/or retention of the genetic capability of producing bacteriocin(s) as the need arises.

4.1.4 Detection of Bacteriocins of Gram-Positive Bacteria

Two simple, agar culture-based methods have been most commonly used for the detection of bacteriocin production in vitro – simultaneous and deferred antagonism. However, some bacteriocins such as streptocins STH_1 and STH_2 appear to be produced only in liquid media (Schlegel and Slade 1973; Tompkins et al. 1997). In simultaneous antagonism, the test and indicator bacteria are typically grown together on an agar surface, and detection of bacteriocin production is dependent on the release of the inhibitory agent(s) relatively early in the growth of the test culture (i.e., before overgrowth of the indicator bacterium). By contrast, the deferred antagonism test, which is most commonly used in bacteriocin typing procedures, allows for independent variation of the incubation parameters (time, temperature, atmosphere) of the test and indicator bacteria. In practice, it is important to test by both methods when screening bacteria for bacteriocinogenicity. Optimal conditions for test strain growth do not necessarily coincide with optimal bacteriocin production conditions. Indeed, bacteriocin production can be enhanced when the producer cells are relatively stressed (nutritionally or environmentally). Specific medium supplements shown to markedly affect the production of bacteriocins by Gram-positive bacteria have included yeast extract (enhancing mutacin production), glucose (effecting catabolite repression of some streptococcal bacteriocins), and magnesium ions (repressing expression of some lantibiotics).

Screening tests for inter-bacterial inhibition on agar media do not, of course, distinguish the activities of bacteriocins from inhibition due to non-bacteriocin agents such as bacteriophage, primary metabolites such as H_2O_2 and lactic acid, or non-ribosomally synthesized antibiotics such as bacitracin. Nor can such tests discriminate inhibition attributable to nutrient depletion or to the combined activities of multiple bacteriocins and/or other inhibitory agents. We recommend the use of the acronym BLIS (for bacteriocin-like inhibitory substance) to refer to as yet uncharacterized inhibitory agents that appear "bacteriocin-like" in their activity; e.g., *St. aureus* BLIS H12 is the initial descriptor for bacteriocin-like activity found to be produced by *St. aureus* strain H12.

4.1.5 Nomenclature of Bacteriocins of Gram-Positive Bacteria

Assignment of a specific bacteriocin designation to an inhibitory agent should occur only after isolation, purification and sequencing of the peptide(s) and of the corresponding bacteriocin structural gene(s), and confirmation of the uniqueness of the active bacteriocin molecule(s) by reference to publicly available sequence databases. Over the years, the naming of bacteriocins of Gram-positive bacteria has been haphazard, being based sometimes upon the species and, at other times, upon the genus of the primary producer strain. Although, in theory, the first described example of a bacteriocin should be accorded naming priority, this is not always adhered to in practice. For example, the name "macedocin" was ascribed to a lantibiotic produced by a strain of *Streptococcus macedonicus* (Georgalaki et al. 2002), and this despite it being identical to the previously described SA-FF22 from *Streptococcus pyogenes* (Jack et al. 1994). Whereas for the colicins (with generally only one type produced per strain) the basis for subdivisions was receptor specificity, as defined by using specific colicin-resistant mutants, it has been more problematic to obtain bacteriocin receptor mutants in Gram-positive bacteria. Furthermore, strains producing multiple bacteriocins have created difficulties in several Gram-positive species, and in our experience this is particularly so for *Streptococcus mutans*, *Streptococcus salivarius*, and *Streptococcus uberis*. Bacteriocins that have only minor conservative differences in the amino acid sequences of their propeptide components, resulting in no significant change to (1) their secondary structure, (2) their activity spectra, and (3) the cross-specificity of their producer strain immunity, should be referred to as natural variants. For example, nisins Z, Q and U are natural variants of the first described nisin A. Similarly, the subsequently described variants of the *S. salivarius* lantibiotic salivaricin A, all of which exhibit differences in their propeptide sequences, have been named salivaricins A1, A2, A3, A4, and A5. All of the salivaricin A variants exhibit the same (putative) bridge pattern structure, cross-inducibility and cross-immunity characteristics. By contrast, the *S. mutans* bacteriocins successively characterized and named by the Caufield group include both lantibiotics (mutacins I, II, III; Chikindas et al. 1995; Qi et al. 1999, 2000) and a non-lantibiotic bacteriocin (mutacin IV; Qi et al. 2001).

4.1.6 Classification of Bacteriocins of Gram-Positive Bacteria

In 1993, Klaenhammer attempted to put some order into the classification of the bacteriocins of LAB, by proposing four major classes (Klaenhammer 1993):

- Class I – post-translationally modified bacteriocins, i.e., the lantibiotics,
- Class II – small (<10 kDa) heat-stable membrane-active bacteriocins,
- Class III – larger (>30 kDa) heat-labile bacteriocins, and
- Class IV – complex bacteriocins composed of essential lipid or carbohydrate moieties in addition to protein.

Class II was further subdivided into IIa (anti-listerial peptides having the amino acid motif YGNGV/L in the N-terminal part of the peptide), IIb (two-component peptides), and IIc (thiol-activated peptides requiring reduced cysteine residues for activity).

This provisional scheme was adopted by most investigators in the field, although it was repeatedly noted that sustainable evidence was lacking for bacteriocins fulfilling the criteria for group IV. The lantibiotics have been generally divided into linear (type A) and globular (type B) subtypes, though additional subdivisions have also been mooted. The class II bacteriocins have occasionally been carved into additional subgroups either for convenience or because of the personal bias of some investigators. Class III, the small group of large cell wall-active bacteriocins, has typically either been dismissed or largely overlooked by those focused on the membrane-active small peptides. More recently, Kemperman et al. (2003a) recommended recognition of a new group (class V) comprising ribosomally synthesized, non-modified head-to-tail ligated cyclic antibacterial peptides.

As standard practice in this laboratory, we first test LAB for their production of bacteriocins by use of a three-step screening process:

1. deferred antagonism bacteriocin "fingerprinting", using a set of nine standard indicator strains (Tagg and Bannister 1979),
2. repeating the bacteriocin fingerprinting procedure, but incorporating a heating step (80 °C for 45 min) prior to application of the indicator (detector) bacteria, and
3. polymerase chain reaction (PCR)-based detection of lantibiotic processing genes (*lanM, lanB* and *lanC*).

This process can sometimes provide preliminary evidence for the production of multiple bacteriocins by the test strain, and also may hint to the possible class of inhibitory molecule(s) being produced. For example, the lantibiotics (class I) typically produce heat-stable inhibition of the *Micrococcus luteus* indicator strain, whereas inhibitory activity due to class III (large) bacteriocins is usually eliminated by the heating step. Our application of these procedures to many different species of LAB has shown that even use of only a single set of nine indicator strains can demonstrate a very high frequency of BLIS detection. *S. mutans, S. salivarius* and *S. uberis* exhibit a particularly high incidence of bacteriocinogenicity, with some strains producing combinations of bacteriocins belonging to different classes. Some notable examples include (1) bacteriocin-producing *S. salivarius* that harbor mega-plasmids (typically 160–220 kb), some of which encode no less than five different bacteriocins (Wescombe et al. 2006a), (2) *S. uberis* 42 that produces the lantibiotic nisin U and uberolysin, a circular (cyclic) bacteriocin (R.E. Wirawan et al. submitted), and (3) *S. mutans* UA140 that elaborates a lantibiotic (mutacin I) and a class II inhibitory agent (mutacin IV; Qi et al. 2001).

Cotter et al. (2005b) have recently proposed a more radical modification to the Klaenhammer classification scheme for LAB bacteriocins, in which there are essentially only two principal categories: lantibiotics (class I) and non-lanthionine-containing bacteriocins (class II). The former class III (large heat-labile murein hydrolases) are renamed bacteriolysins, and class IV (the lipid- or carbohydrate-containing bacteriocins) is withdrawn. It was further suggested that the Klaenhammer class II subgroups IIa (listeria-active peptides) and IIb (two-peptide bacteriocins) be retained, and that the class V (cyclic peptides) proposed by Kemperman et al. (2003a) be reassigned as class IIc. Class IId was proposed to be a repository for all of the remaining linear non-lanthionine-containing bacteriocins. It should be noted that Cotter et al. commented only upon those class III members that possess a lytic mode of action (e.g., lysostaphin), despite reports of non-lytic large heat-labile bacteriocins such as helveticin J (Joerger and Klaenhammer 1986) and streptococcin G-2580 (Tagg and Wong 1983). Although we concur with some elements of the revised classification scheme of Cotter et al. (2005b) for LAB bacteriocins, some of our own experiences with these molecules lead us to propose some further modifications for consideration (outlined in Fig. 4.1), which can generally be applicable to all bacteriocins from Gram-positive sources:

- In agreement with Cotter et al.:
  1. Klaenhammer's class IV (chemically complex bacteriocins) is eliminated, and
  2. Class IIc (thiol-activated peptides) is eliminated.
- In contrast to Cotter et al.:
  1. Class III (large bacteriocins) is retained, and is now subdivided into IIIa (bacteriolysins) and IIIb (non-lytic proteins),
  2. the cyclic bacteriocins (class IIc) now constitute a newly defined class IV, and
  3. as a consequence of upgrading the cyclic bacteriocins to their own class, type IIc now becomes the repository for all unmodified class II inhibitors other than the listeria-active (type IIa) and multi-component bacteriocins (type IIb).

Clearly, research in the field of LAB bacteriocins is still progressing exponentially, and it is not easy to formulate an enduring natural classification scheme that encompasses all of the existing bacteriocin-like proteins (the evolutionary origins of which appear quite independent) as well as potentially accommodating as yet unimagined novel members.

In this chapter, we attempt to provide the reader with an overview of some of the diversity of bacteriocins elaborated by Gram-positive bacteria, admittedly heavily biasing our attention toward the LAB, where much contemporary research in this field has been conducted. Wherever possible, the most recent developments within an existing class of bacteriocins, and information relating to previously undescribed classes of inhibitors will be given greatest emphasis.

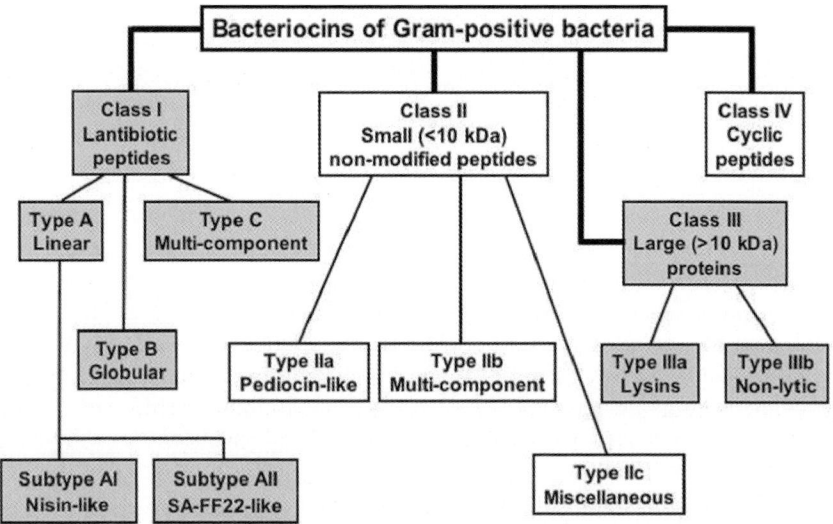

Fig. 4.1 Our proposed classification schema, based on that of Cotter et al. (2005b), but modified so as to be applicable to most, if not all, known bacteriocins of Gram-positive bacteria

4.2 Class I: The Lanthionine-Containing (Lantibiotic) Bacteriocins

The term lantibiotic (Schnell et al. 1988; Jung 1991) refers to the diverse array of bacterial antibiotic peptides that contain the non-genetically encoded amino acids lanthionine (Lan) and/or 3-methyllanthionine (MeLan), as well as various other highly modified amino acids, commonly including the 2,3-unsaturated amino acids dehydroalanine (Dha) and dehydrobutyrine (Dhb). All of the lantibiotics currently described are thought to be produced as ribosomally synthesized precursor peptides, which then undergo a series of post-translational modification reactions to produce the unusual amino acids described above as intrinsic components of the biologically active peptides. To date, lantibiotics have been found to be produced only by Gram-positive bacteria. Furthermore, they are generally considered to predominantly, if not exclusively, act on Gram-positive targets. As the family of lantibiotic molecules grew, the individual members were initially classified according to the topology of their ring structures and their biological activities (Jung and Sahl 1991), as either type A (elongated amphipathic structures) or type B (globular and more compact structures). In order to encompass the more recently described two-component varieties, the type C lantibiotics has been proposed. The type A lantibiotics are further divided into subtypes AI and AII based on the size, charge and sequence of their leader peptides (de Vos et al. 1995). It must be noted, however, that the lantibiotics are a difficult group to subdivide, and indeed it has been proposed that on the basis of similarities in their unmodified propeptide sequences, they could be split into 11 groups (Cotter et al. 2005a). The lantibiotics have been reviewed extensively over the last decade, and the reader is referred to some of these accounts for a more complete overview (McAuliffe et al. 2001b; Chatterjee et al. 2005). We propose to focus only on a selection of lantibiotics that we consider illustrate some of the significant diversity of these molecules, and to update the reader on recent developments in the field not covered by other reviews (cf. Table 4.1).

4.2.1 Type AI Lantibiotics

The prototype type AI lantibiotic, nisin, is perhaps the most extensively characterized of all bacteriocins. Produced by *Lactococcus lactis*, nisin has a long history of research, its discovery in 1928 (Rogers 1928) predating that of penicillin. Nisin has now been used safely in the food industry as a preservative for over 40 years without the appearance of significant bacterial resistance. Since the nisin biosynthetic pathway, requiring the coordinated expression and action of at least 11 gene products, is generally mirrored by most other lantibiotics, we will give a brief description of the processes involved. The precursor peptide, encoded by *nisA*, is acted upon by the proteins NisB and

Table 4.1 A selection of class I (lantibiotic) bacteriocins

Bacteriocin	Producer species	Distinctive characteristics	Recommended reference(s)
Type A			
Subtype AI			
Nisin A	*Lactococcus lactis*	Prototype AI lantibiotic, widely used as a food preservative for over 40 years; the best-characterized of the bacteriocins of Gram-positive bacteria; induces its own expression through binding to a two-component signal transduction system	Mattick and Hirsch (1947), Gross et al. (1971), McAuliffe et al. (2001b)
Nisin U	*Streptococcus uberis*	First nisin variant to be characterized from a non-lactococcal host bacterium	Wirawan et al. (2006)
Streptin	*Streptococcus pyogenes*	Utilizes a general bacterial protease (SpeB), rather than a dedicated protease (LanP) to remove the leader sequence from the prepeptide	Wescombe and Tagg (2003)
Subtype AII			
SA-FF22	*S. pyogenes*	First subtype AII lantibiotic to be characterized; has a two-component sensor-kinase histidine-response regulator system for which the inducing molecule (most likely a peptide) is not SA-FF22 itself	Jack et al. (1994)
Lacticin 481	*L. lactis*	Considered by some to be the prototype of this subtype; in contrast to SA-FF22, the lacticin 481 locus does not encode a two-component sensor-kinase histidine-response regulator	van den Hooven et al. (1996)
Salivaricin A	*Streptococcus salivarius*	Multiple SalA variants have been characterized from different species of streptococci; all tested M serotypes of *S. pyogenes* (other than M11 and M37) have the SalA structural gene variant *salA1*; each SalA variant has been shown to effect both auto- and heterologous induction of SalA production in their respective host strains	Ross et al. (1993), Wescombe et al. (2006b)

Sublancin 168	*Bacillus subtilis*	The only lantibiotic shown to have a disulfide bond in addition to the typical MeLan and Dha residues	Paik et al. (1998)
Type B			
Mersacidin	*Bacillus* spp.	Prototype type B lantibiotic, and the only lantibiotic of which the structure has been resolved by X-ray crystallography; effective in treating systemic staphylococcal (*St. aureus*) infections and nasal carriage in a mouse model system	Chatterjee et al. (1992)
Cinnamycin	*Streptomyces cinnamoneus*	The only characterized lantibiotic thought to be exported by the general secretory (Sec) pathway; has two MeLan rings in which the nucleophilic Cys is positioned N-terminally to the Dhb; may have use as an anti-inflammatory and anti-allergy drug	Kessler et al. (1987)
Type C			
Lacticin 3147	*L. lactis*	The prototype two-component lantibiotic; structures and functions of both peptide components have been defined; killing of sensitive cells requires initial binding of LtnA1 to lipid II, followed by complexing with LtnA2 to effect pore formation	Martin et al. (2004)
Smb	*Streptococcus mutans*	Regulated in response to the external levels of CSP; a variant (BHT-A) has been identified in *S. rattus* that appears to be closely associated with the locus for the class II bacteriocin BHT-B	Yonezawa and Kuramitsu (2005), Hyink et al. (2005)
Cytolysin	*Enterococcus faecalis*	A lantibiotic that exhibits toxicity for both bacterial and eukaryotic cells; cytolysin subunits are activated by a two-step process involving transport and initial GG specific cleavage by CylB (an ABC transporter), following which the cytolysin subunits are activated via the removal of an additional six amino acids by CylA (a serine protease)	Booth et al. (1996), Coburn and Gilmore (2003), Coburn et al. (2004)

NisC to dehydrate particular Ser/Thr residues, some of which are then used to form specific thioether bonds (i.e., Lan and MeLan) with Cys residues located (generally) further toward the C-terminus of the molecule. NisT is an ABC transporter responsible for export of the modified prepeptide, and NisP is a membrane-anchored protease able to cleave the leader peptide to release active nisin. NisI is involved in nisin immunity by an as yet ill-defined mechanism. Although the majority of NisI appears to be localized within the cytoplasmic membrane of the producer cell (Qiao et al. 1995), a significant amount is also secreted into the cytoplasm where it may bind to external nisin before it can aggregate at the cell surface (Koponen et al. 2004). Expression of *nisI* in nisin-sensitive *L. lactis* strains results in moderately decreased sensitivity to nisin (Qiao et al. 1995). However, full immunity levels are not achieved without the presence of NisFEG. NisFEG is an ABC transport protein complex, presumably contributing to nisin immunity in a manner similar to that used by multi-drug transporters, by reducing the concentration of nisin in direct contact with the cytoplasmic membrane. NisR and NisK together form a two-component sensor-kinase/response-regulator element involved in the regulation of nisin biosynthesis, which characteristically occurs late in the exponential phase of growth. Interestingly, since nisin itself is the specific ligand recognized by the sensor NisK, it up-regulates its own expression (Kuipers et al. 1995). The basic elements of the nisin biosynthetic pathway are conserved for all lantibiotics, with only minor variations such as the use of the LanM modification enzyme, rather than of the LanB/C complex for dehydratase and ring formation reactions and the encoding of LanD enzymes by a minority of lantibiotic loci to effect formation of the unusual amino acids S-[(Z)-2-aminovinyl]-D-cysteine (AviCys) and S-[(Z)-2-aminovinyl]-(3S)-3-methyl-D-cysteine (AviMeCys). For some lantibiotics, specific immunity appears attributable either only to LanI (e.g., Pep5; Reis et al. 1994) or only to the LanFEG system (e.g., lacticin 481; Rince et al. 1997). On the other hand, the epidermin and gallidermin gene clusters encode an additional accessory factor LanH, which enhances LanFEG-mediated immunity (Hille et al. 2001).

In addition to its widespread use as a food preservative, nisin and other members of the lantibiotic class have been investigated for their potential applications in medicine. The MICs of mutacin B-Ny266 and nisin A were shown to be comparable to those of vancomycin and oxacillin against various bacterial pathogens (Mota-Meira et al. 2000). Both lantibiotics were active against vancomycin- and oxacillin-resistant strains of *Helicobacter pylori* and *Neisseria* spp., making them potential candidates for treatment of infections caused by these bacteria (Hancock 1997; Mota-Meira et al. 2000). A novel potential application of nisin is as a spermicidal contraceptive. Studies with rabbits indicated that vaginal administration of 1 mg of nisin stopped sperm motility completely, none of the treated animals having become pregnant (Reddy et al. 2004). Complete histopathologic evaluation of the vagina indicated no adverse effects resulting from the intravaginal application of nisin, in terms of either tissue damage or subsequent reproductive performance (Aranha et al. 2004; Reddy et al. 2004). A future direction for lantibiotic application may involve the rational design of new peptides based on desirable structural features of some well-characterized biologically active peptides such as nisin. An analysis of 37 known lantibiotics indicated that although there were no hard and fast rules, Ser/Thr residues were more likely to be dehydrated when flanked by hydrophobic amino acids than by hydrophilic residues. To test the predicted dehydration sequence rules, hexapeptide-encoding sequences were fused to the nisin leader peptide, and expressed in a *L. lactis* strain containing the nisin modification and export enzymes. Analysis of the composition of the hexapeptide products confirmed the designers' predictions, demonstrating the feasibility of rational design of novel peptides having specific dehydrated amino acid residues (Rink et al. 2005).

As ever more lantibiotics are being detected, it has become increasingly obvious that a continuum of natural variants exists, some exhibiting only a single amino acid residue difference from previously documented lantibiotics, but others having multiple sequence variations. A variant of nisin produced by *Streptococcus uberis* strain 42 has recently been identified (Wirawan et al. 2006), the first of the nisin family not produced by a *Lactococcus* strain. The biologically active 31-amino acid (aa) nisin U differs from the 34-aa nisin A in 12 of its amino acids (82% similarity of the propeptides; Fig. 4.2a). Nisin U is predicted to share the same bridging pattern as nisin A, and the producer strains of nisin A and nisin U are cross-immune. This apparent cross-immunity to the two nisin peptides is particularly interesting, since the putative immunity peptide for nisin U, NsuI, shares only 55% homology with NisI. By contrast, there is no indication of cross-immunity between subtilin (from *Bacillus subtilis*) and nisin, despite them having 60% propeptide sequence similarity (McAuliffe et al. 2001b). In addition, the antimicrobial spectrum of nisin U appears to match closely that of nisin A, although there appears to be some relative reduction in the activity of nisin U against *S. pyogenes* and *L. lactis* strains. Significantly, nisin U and nisin A exhibited both auto-inducing and cross-inducing activity when added to cultures of the respective nisin-producing *Lactococcus* and *Streptococcus* strains, further emphasizing the close functional identity of the two peptides and justifying the classification of the *S. uberis* lantibiotic as a nisin variant. The nisin U genetic locus comprises 11 open reading frames, closely similar to their nisin A counterparts, but with *nsuPRKFEG* located upstream of *nsuA* rather than downstream of *nsuI*, as in the nisin A locus (Fig. 4.2b). The nisin U locus is flanked by transposon-related sequences, and also has a 742-bp region between *nsuG* and *nsuA* encoding remnants of a transposase (Fig. 4.2b), indicating that a rearrangement of the locus has occurred. Streptin is a 23-aa type AI lantibiotic produced by *S. pyogenes* strain M25 exhibiting similarity in its first two ring structures with the corresponding region in nisin (Karaya et al. 2001; Wescombe and Tagg 2003). The streptin locus appears similar to that of subtilin, in that it does not encode a specific protease (LanP) for propeptide

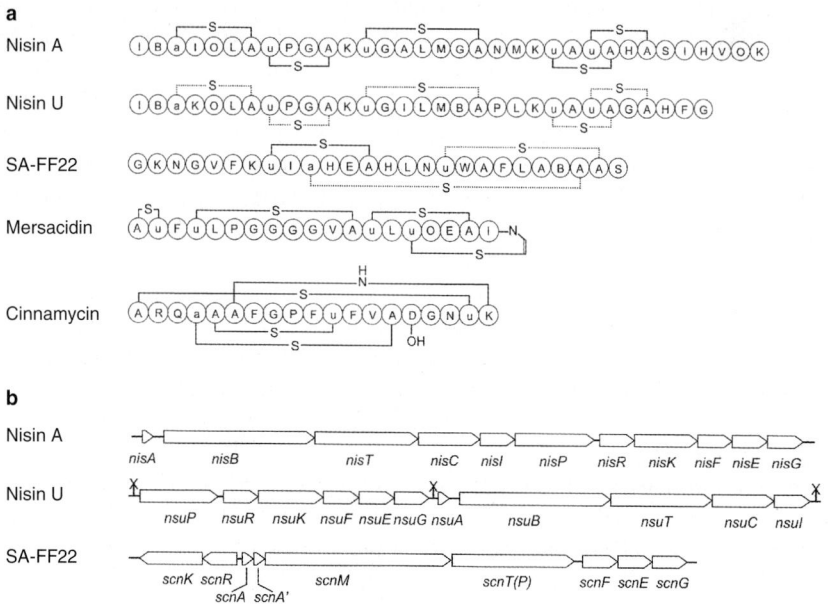

Fig. 4.2 a The primary structures of the type A lantibiotics nisin, nisin U and SAFF-22, and the representative type B lantibiotics mersacidin and cinnamycin. Modified amino acid abbreviations: a, D-alanine; B, 2,3-dihydrobutyrine; O, 2,3-dihydroalanine; u, D-α-aminobutyric acid; a-S-A, lanthionine (Lan); u-S-A and A-S-u, methyllanthionine (MeLan); A-NH-K, lysinoalanine; D-OH, *erythro*-3-hydroxyaspartic acid. All other conventional amino acids are given in one-letter code. The *solid lines* represent the Lan and MeLan bridges that have been confirmed experimentally, whereas those with *dotted lines* are predicted. **b** Organization of the biosynthetic loci of nisin A, nisin U and SA-FF22. Note the different order of the *lanPRKFEG* genes between the nisin A and nisin U loci. The *X symbols* in the nisin U locus represent remnants of mobile genetic elements (see text)

activation (Stein and Entian 2002). Rather, it appears that the producers of subtilin and streptin utilize other host cell proteases to remove the lantibiotic leader sequences, probably following prepeptide export. In the case of streptin, the *S. pyogenes* cysteine proteinase, SpeB, has been implicated in the prepeptide cleavage reaction, since proteinase-negative mutants of strain M25 concomitantly lose the ability to express the streptin phenotype (Hynes and Tagg 1986; S. O'Brien and J.R. Tagg, unpublished data). The utilization of host cell proteases for the processing of lantibiotics could be viewed as an efficient way to reduce the metabolic burden of lantibiotic production, although it may limit the dissemination of the locus to other species.

4.2.2 Type AII Lantibiotics

The type AII lantibiotics differ from those in type AI in that their thioether ring formation is effected by bifunctional LanM enzymes, rather than by the combined action of LanB and LanC, and also because they generally have the conserved consensus sequences E(L/V)S and E(L/M) in their leader peptides. Furthermore, their leader sequences resemble more closely those of class II bacteriocins, in that they contain a "double-glycine" (GG/GA/GS) motif immediately preceding the cleavage site (McAuliffe et al. 2001a; Chatterjee et al. 2005). This group of lantibiotics also includes a most unusual member, sublancin 168 produced by a *B. subtilis* strain, which appears to be the first bacteriocin to contain both lanthionine ring structures and stabilizing disulfide bonds (Paik et al. 1998).

Although lacticin 481 is largely touted as the prototype of this subclass, the first of this group to be characterized was actually the 26-aa lantibiotic streptococcin A-FF22 (SA-FF22) produced by *S. pyogenes* strain FF22 (Jack and Tagg 1991, 1992). Lacticin 481, now extensively characterized, is a 27-aa lantibiotic containing two Lan, one MeLan and one Dhb residue (Piard et al. 1992). Interestingly, the lacticin 481 genetic locus, unlike that of most type AI lantibiotics, appears not to encode a two-component sensor-kinase response-regulator system, rather being regulated at the transcriptional level by pH control of P1 and P3 promoters located upstream of the structural gene *lctA* (Hindre et al. 2004). By contrast, the locus encoding SA-FF22 in *S. pyogenes* does have a two-component sensor-kinase response-regulator system, but this responds not to the inhibitory lantibiotic SA-FF22 but to another putative signal molecule (P.A. Wescombe, unpublished data). In fact, the molecular mechanisms of lantibiotic regulation are strikingly diverse, with examples of

1. negative regulation of members of the type AII (e.g., lactocin S; Rawlinson et al. 2002) and two-component lantibiotics (e.g., lacticin 3147, McAuliffe et al. 2001a; cytolysin, Haas et al. 2002),
2. no genes encoding regulatory elements within the locus (e.g., lacticin 481; Hindre et al. 2004),
3. homologous (auto) regulation (e.g., nisin, Kuipers et al. 1995; salivaricin A, Upton et al. 2001),
4. heterologous regulation using a signal peptide differing from the induced lantibiotic (e.g., SA-FF22; P.A. Wescombe, unpublished data), and
5. regulation by two peptides, each influencing expression of different genes within the locus (for example, the mersacidin locus encodes the response regulators MrsR1 and MrsR2, where MrsR1 regulates immunity gene expression and MrsR2 regulates lantibiotic biosynthesis; Guder et al. 2002).

In our laboratory, we have conducted extensive research on salivaricin A (SalA), a 22-aa type AII lantibiotic produced by *S. salivarius* (Ross et al. 1993). Five closely related variants of SalA have recently been described (Wescombe et al. 2006b), each shown to effect both auto- and heterologous induction of SalA production in the respective host strains. The novelty of SalA lies in the wide distribution of the SalA (and variant) locus in *Streptococcus* species, having now been detected in *S. salivarius*, *S. pyogenes*,

Streptococcus dygalactiae and *Streptococcus agalactiae*. Oddly, the structural gene *salA1* was detected in *S. pyogenes* of 65 different M serotypes (Simpson et al. 1995). Only two strains (of serotypes 11 and 37) did not harbor *salA1*. At first glance, this appears to be anomalous, since the majority of *S. pyogenes* are inhibited by SalA when tested in vitro (Ross et al. 1993). However, it has now been demonstrated that, other than in serotype M4T4 *S. pyogenes*, all SalA1 loci are non-functional, due (at least in part) to either deletions in the genes encoding SalM and SalT, or frameshift mutations in the *salT* gene (Wescombe et al. 2006b). It is tempting to speculate that this very common retention, especially of the immunity-associated components of the SalA locus in *S. pyogenes*, may be ecologically driven, as both *S. pyogenes* and *S. salivarius* are inhabitants of the human oral cavity. The *S. pyogenes* serotype M11 and M37 prototype strains are unusual in that they do not possess the SalA immunity genes (P.A. Wescombe et al., unpublished data). The M11 strain is an A-variant *S. pyogenes*, thought to have lost the ability to assemble intact group A carbohydrate during the course of prolonged serial subculture in vitro (D. Johnson, personal communication). The lack of an obvious selective advantage associated with SalA immunity for *S. pyogenes* strains grown for prolonged periods as laboratory monocultures could favor the loss of immunity-related components of the locus. The M37 prototype strain is also very unusual, in that no other examples of strains of this serotype appear to have been isolated (D. Johnson, personal communication). Both of these observations are consistent with a survival advantage for *S. pyogenes* in situ being linked to their retention of at least the immunity-related components of the *salA* locus. Hence, in this case, the acquisition and retention of lantibiotic genetic elements may have contributed to the adaptation and survival of a bacterial species.

It has recently been observed that the SalA locus is borne on large (>150 kb) plasmids in *S. salivarius*, whereas in *S. pyogenes* the locus appears typically to be chromosomally located. The ubiquitous presence of *salA* in *S. pyogenes* indicates that the acquisition of this locus was an early event in the establishment of the species, or at least that only strains of *S. pyogenes* that are capable of expressing SalA immunity have maintained associations with the human host. The large plasmids in bacteriocin-producing *S. salivarius* have been found to harbor loci for various combinations of streptococcal lantibiotics including salivaricin A, salivaricin B, streptin, and a variant of SA-FF22. The lantibiotic loci appear to be juxtaposed in contiguous segments, separated by no more than ca. 4 kb of non-lantibiotic-related DNA. Moreover, genes encoding Tra-like proteins (potentially involved in conjugative transfer of the plasmids) have also been identified. These observations support the hypothesis that cassettes of lantibiotic loci could be disseminated together, thereby rapidly expanding the antimicrobial arsenal of the recipient strain. Indeed, in vivo transfer of the entire 180 kb of *S. salivarius* K12 bacteriocin-associated plasmid to indigenous *S. salivarius* has been demonstrated to occur in the oral cavity of subjects colonized with the probiotic

S. *salivarius* K12 (Wescombe et al. 2006a). The wide distribution of closely related lantibiotic loci throughout different oral streptococcal species indicates a high frequency of horizontal gene transfer. In the case of *S. salivarius*, the large plasmids appear to have been particularly effective at acquiring additional bacteriocin loci, and our preliminary findings indicate that most BLIS-producing *S. salivarius* strains have plasmids of size > 40 kb.

4.2.3 Type B (Globular) Lantibiotics

The type B lantibiotics are more globular and compact in shape than those of type A, and generally are either uncharged or negatively charged at neutral pH. Mersacidin, the prototype for this group, is a 20-aa peptide (mass 1,825 Da) and its distinctive features include three MeLan rings, one Dha and a S-[(Z)-2-aminovinyl]-(3S)-3-methyl-D-cysteine (AviMeCys) residue (Chatterjee et al. 1992). Mersacidin, which derives its name from its potent activity against methicillin-resistant *St. aureus* (MRSA, the hospital-acquired "superbug"), is also the only lantibiotic of which the structure has been resolved by X-ray crystallography (Schneider et al. 2000). Mersacidin does not form pores in bacterial membranes, but rather inhibits peptidoglycan synthesis through a specific interaction with the peptidoglycan precursor lipid II (Brotz et al. 1997). The sequestering of lipid II prevents its utilization by the transpeptidase and transglycosylase enzymes that install the crosslinked network of the bacterial cell wall. Both nisin and mersacidin appear to bind to a different portion of lipid II than does vancomycin (the antimicrobial of last resort for the treatment of multiply antibiotic-resistant *St. aureus*), indicating that these molecules may prove to have important chemotherapeutic applications (Brotz et al. 1995; Breukink et al. 1999). Indeed, mersacidin has been shown to be very effective for the treatment of systemic staphylococcal infections, and in eliminating nasal carriage of *St. aureus* in a mouse model system (Chatterjee et al. 1992; Kruszewska et al. 2004). Lipid II, however, does not serve as a docking molecule for all lantibiotics, since Pep5 and epilancin K7 have been shown specifically not to bind lipid II. These molecules presumably have an alternative docking molecule or receptor, since they have greater activity than other pore-forming molecules against certain indicator bacteria (Brotz et al. 1998; Pag et al. 1999).

Cinnamycin is a 19-aa type B lantibiotic produced by *Streptomyces cinnamoneus*, and has also been purified as Ro 09-0918 (Kessler et al. 1987) and lanthiopeptin (Naruse et al. 1989; Palmer et al. 1989). Its structure is novel in that it has two MeLan residues in which the nucleophilic cysteine is positioned N-terminally to the Dhb. Although also found in some other type B lantibiotics, this direction of ring formation has so far not been observed in any of the type A lantibiotics other than the LtnA1 peptide of the lacticin 3147 two-component lantibiotic system. Another unusual feature of cinnamycin is a head-to-tail lysinoalanine bridge, the formation of which has so far not been ascribed to any particular gene product in the comprehensive array of putative ORFs identified in the cinnamycin locus. Intriguingly, the CinA prepeptide has a much longer leader peptide than that of other lantibiotics, and it has been proposed that cinnamycin may be secreted by a more general export mechanism such as the general secretory (Sec) pathway, once again illustrating the broad diversity of the lantibiotic class (Widdick et al. 2003). Cinnamycin has been shown to be a potent inhibitor of phospholipase A2 (an enzyme involved in the synthesis of prostaglandins and leukotrienes in the human immune system) through the sequestration of its substrate phosphatidylethanolamine. Due to this activity, cinnamycin may prove to have a useful application as an anti-inflammatory and anti-allergy drug (Marki et al. 1991).

4.2.4 Type C (Multi-Component) Lantibiotics

Each of the multi-component lantibiotic consortia described to date comprise two post-translationally modified peptides that individually have little or no activity, but display strong synergistic antibacterial action. Lacticin 3147 produced by *Lactococcus lactis* DPC3147 is arguably the most intensively studied member of this group, and both component peptides (LtnA1 and LtnA2) have been structurally characterized (Ryan et al. 1999). Features of the peptides include Lan and MeLan residues, and also a D-Ala residue derived from L-Ser by post-translational modification. Interestingly, the structure of LtnA1 bears some resemblance to that of the type B lantibiotic mersacidin, and the LtnA2 peptide displays some similarity to lactocin S (a type AII lantibiotic). The obvious structural differences between LtnA1 and LtnA2 therefore require that the genetic locus for lacticin 3147 encode two LanM enzymes, each of which is presumably responsible for the post-translational modification of one of the peptides (McAuliffe et al. 2000). The mechanism of action of lacticin 3147 has recently been shown to result from the sequential action of the two peptides, on condition that LtnA1 be added prior to LtnA2 (Morgan et al. 2005). It was therefore inferred that LtnA1 binds lipid II (a reaction responsible for the independent inhibitory activity displayed by LtnA1), following which LtnA2 interacts with the LtnA1–lipid II complex to bring about more effective insertion into the target membrane and pore formation, with an associated 30-fold increase in inhibitory activity compared to that obtained by LtnA1 alone (Morgan et al. 2005).

Smb (*Streptococcus mutans* bacteriocin) was recently shown to be a two-component lantibiotic (Yonezawa and Kuramitsu 2005). Expression of SmbA and SmbB by the *smb* locus appears to be regulated in response to the external levels of a competence-stimulating peptide (the peptide that activates the development of competence for genetic transformation; see below). It is possible that the production of the lantibiotic has been coupled to the competence cascade to ensure that there is an abundance of exogenous DNA available for uptake by the newly competent bacteria. Alternatively, it may be that the apparent co-regulation is purely a consequence of the insertion of the lantibiotic locus into a region having the competence promoter upstream. A variant of the Smb lantibiotic, named BHT-A, was recently identified in *Streptococcus rattus* strain BHT, and shown to be composed of the two peptides BHT-Aα and BHT-Aβ (Hyink et al. 2005). Interestingly, the Smb/BHT-A locus appears to be closely linked to the locus for BHT-B, a class II bacteriocin, in all *S. rattus* strains examined to date (Hyink et al. 2005).

Cytolysin, produced by *Enterococcus faecalis*, consists of two lantibiotic subunits (CylL$_L$ and CylL$_S$), and is the only lantibiotic confirmed to exhibit toxicity for both bacterial and eukaryotic cells. Although not all strains of *E. faecalis* are hemolytic, the occurrence of hemolysis is higher among clinical isolates, especially those from the bloodstream (Booth et al. 1996). As many as 60% of infection-associated *E. faecalis* elaborate cytolysin, and it has been shown to lower the LD$_{50}$ of *E. faecalis* for mice and to contribute to toxicity in experimental endocarditis and endophthalmitis models. In addition, cytolysin-positive strains are associated with a fivefold increased risk of acute terminal outcome in patients with nosocomial enterococcal bacteremia (Coburn et al. 1999). Interestingly, cytolysin is encoded on large pheromone-responsive conjugative plasmids, which may, at least in part, account for the high prevalence of the cytolysin locus in enterococci.

The CylL$_L$ and CylL$_S$ subunits are activated by a two-step process involving initial transport and GG site-specific cleavage (to CylL$_L$' and CylL$_S$') mediated by the ABC transporter CylB, followed by removal of a further six amino acids (forming CylL$_L$" and CylL$_S$") by the serine protease CylA. It was shown that in order for the cytolysin to efficiently lyse erythrocytes and bacterial cells, both subunits need to be fully processed by CylA (Booth et al. 1996).

Expression of the cytolysin locus is directly regulated by the synergistic action of two repressor proteins CylR1 and CylR2, both of which lack homologues of known function (Haas et al. 2002). De-repression occurs at a specific cell density when one of the cytolysin subunits (CylL$_S$") reaches an extracellular threshold concentration. These observations form the basis for a model of cytolysin auto-induction by a quorum-sensing mechanism involving a novel two-component regulatory system (Haas et al. 2002). CylL$_L$ and CylL$_S$ expression is further regulated in response to aerobiosis, with transcription being up-regulated under anaerobic conditions (Day et al. 2003).

Comparison of the cytolysin determinants with those of the type AII lantibiotic lactocin S (from *Lactobacillus sake*) indicates they may share a common ancestry (Gilmore et al. 1996). Although no cytotoxicity for eukaryotic cells has been reported for lactocin S, it seems prudent to perform toxicity tests on any lantibiotics (particularly two-component forms) that may have human or veterinary applications to assess their potential for disruption of eukaryotic membranes. Indeed, this revelation of the dual toxicity of cytolysin sounds a timely warning for those contemplating the engineering of novel lantibiotics, since it demonstrates the potential for these molecules to exhibit toxicity for eukaryotic cells, perhaps sometimes by forming multi-component membrane poration complexes in combination with heterologous bacteriocins produced by indigenous bacteria (Wescombe et al. 2005).

4.3 Class II: The Unmodified Peptide Bacteriocins

Class II essentially encompasses all of the currently described, small (<10 kDa) unmodified (i.e., non-lantibiotic and non-cyclic) peptide bacteriocins of Gram-positive bacteria (Eijsink et al. 2002). As a result, this class comprises over 50 members with diverse origins, ranging from genera inhabiting the oral cavity and gastrointestinal tract (of humans and other animals) to species best known for their involvement in the dairy and food industries. As with the lantibiotics, class II includes inhibitors either functioning as single peptides or requiring the coordinated activity of two or more component peptides. Furthermore, some bacteriocin-like peptides that conform to the class II definition do not appear to possess intrinsic activity of their own, but function to activate bacteriocin biogenesis (Eijsink et al. 2002).

4.3.1 Type IIa: The Pediocin-like Peptides

The largest single collection of class II bacteriocins, at present consisting of over 20 members, is type IIa, the so-called pediocin-like bacteriocins that are epitomized by their particular effectiveness in killing the food-borne pathogen *Listeria monocytogenes* (for recent and comprehensive reviews, refer to Rodriguez et al. 2002 and Fimland et al. 2005). It is this characteristic that has provided the impetus for research of this family of molecules, due to their potential applications as food biopreservatives (Rodriguez et al. 2002). Despite the moniker of this bacteriocin family, it was actually leucocin A from *Leuconostic gelidum* that was the first member to be described (Hastings et al. 1991). Nevertheless, pediocin PA-1, a 44-aa peptide produced by *Pediococcus acidilactici*, is the best-characterized member of this family and therefore justifiably the prototype of this group (Fimland et al. 2005). Moreover, pediocin PA-1 is the only type IIa bacteriocin to be used commercially, i.e., as the active ingredient in Alta™*2341*, an anti-*Listeria* food preservation product (Rodriguez et al. 2002).

Aside from their characteristic anti-listerial activity, the pediocin-like peptides (which vary in length from 37 to 58 residues) are typified by (1) the highly conserved amino acid motif Tyr–Gly–Asn–Gly–Val/Leu (also known as the YGNGV motif or "pediocin box") near their N-termini, and (2) the presence of two cysteine residues forming a disulfide bond. With few exceptions, the members of this group have a similar spectrum of bacterial inhibition, although the MIC values toward the target organisms vary considerably.

In general, the peptides possessing four cysteine residues (forming two disulfide bridges) have a much broader inhibitory spectrum than those with only one disulfide bond (Eijsink et al. 1998). While the conserved YGNGV motif has historically been considered to be the *Listeria*-active part of type IIa bacteriocins, exceptions have been noted, namely acidocin A from *Lactobacillus acidophilus*, which is still anti-listerial despite possessing a slightly altered pediocin box (YGTNGV; Kanatani et al. 1995). In some cases, substitutions within the YGNGV motif strongly reduce the activity of the molecules not only against *Listeria* but also toward other lactic acid bacteria (Miller et al. 1998). Similarly, experiments involving hybrid molecules of four pediocin-like peptides have demonstrated that the C-terminal module of each peptide plays an important role in determining the inhibitory spectrum (Fimland et al. 1996). Therefore, while the pediocin box appears to correlate with anti-*Listeria* activity, the specifics of the biological activity of each type IIa peptide are probably determined by the collective contributions of its amino acid constituents.

The bacteriocidal mode of action of pediocin PA-1 appears to involve three basic steps: (1) binding to the cytoplasmic membrane, (2) insertion of the bacteriocin molecules into the membrane, and (3) formation of the poration complex that permeabilizes the membrane, thereby disrupting the proton motive force and leading to cell death (Rodriguez et al. 2002). Through analyses of bacteriocin-resistant *Lis. monocytogenes* strains, it appears that the receptor for type IIa bacteriocins (including pediocin PA-1) is a mannose-specific phosphotransferase (PTS) system (Ramnath et al. 2000; Gravesen et al. 2002; Vadyvaloo et al. 2004). Although the molecular mechanism of how a pediocin-like peptide interacts with its putative receptor remains to be elucidated, it is possible that the PTS complex acts as a docking molecule that stabilizes the pediocin-mediated pore, not unlike the role of lipid II in the mode of action of nisin.

The genetic determinants for pediocin PA-1 production have been determined to be plasmid-borne in all producing strains examined to date. The pediocin (*ped*) locus consists of four genes *pedA*, *pedB*, *pedC* and *pedD* transcribed as a single polycistronic unit, with *pedA* and *pedB* encoding the pediocin PA-1 prepeptide and its immunity protein, respectively (Venema et al. 1995; Rodriguez et al. 2002). Interestingly, the pediocin PA-1 prepeptide is approximately 80% as active as the mature bacteriocin (Venema et al. 1995). The first 18 amino acids of PedA, the 62-aa pediocin PA-1 precursor, constitute the signal peptide that is cleaved off at the C-terminal side of two glycine residues (the so-called double-glycine or GG motif) during export (Rodriguez et al. 2002; Fimland et al. 2005). This event is carried out by the ATP-binding cassette (ABC) transporter complex (also known as a type I secretion system) composed of the proteins encoded by *pedC* and *pedD*. PedD (724 aa), the actual ABC transporter, is composed of three domains: (1) an N-terminal peptidase domain, which presumably cleaves the prepeptide at the GG motif, (2) a central cell membrane-spanning domain, and (3) a C-terminal

ATP-hydrolyzing domain (Fimland et al. 2005). The role of PedC, the so-called "ABC transporter accessory protein", is not entirely clear, but it may function to aid the passage of the bacteriocin precursor through PedD (Rodriguez et al. 2002).

The core class II bacteriocin-associated genetic locus, which can be found in most unmodified Gram-positive bacteriocin systems irrespective of origin, is essentially composed of the genes encoding

- the bacteriocin precursor peptide (including the GG motif),
- the immunity-associated protein, and
- components of the ABC transporter complex.

Naturally, variations on this theme do exist, such as (1) the organization of the genes, e.g., the ABC transporter gene located upstream of its accessory factor, (2) the absence of a genetic determinant encoding an accessory factor, or (3) the disparate locations of the genes encoding the ABC transporter and the GG-motif-containing peptides. It is noteworthy that the prepeptide of colicin V, an 8-kDa *E. coli* microcin, possesses a GG motif and its biosynthetic locus clearly conforms to the definition of a "core class II locus" (Håvarstein et al. 1994).

An example of the diversity within type IIa is exemplified by the biosynthesis of certain members of the pediocin family such as listeriocin 743A, bacteriocin 31 and enterocin P. These bacteriocins have been shown to be exported via the general secretory (Sec-dependent) pathway, and hence possess Sec-type signal peptides (Cintas et al. 1997; Kalmokoff et al. 2001; Fimland et al. 2005). Such a mechanism of export is potentially beneficial to the host due to its apparent metabolic and genetic economy, i.e., eliminating the need to commit resources to the synthesis of two sets of transport systems.

While most type IIa molecules have been isolated from bacteria usually associated with food products, we have recently identified, from *S. uberis* strain E, a 5.3-kDa anti-listerial bacteriocin possessing a typical pediocin box (G.A. Burtenshaw et al., unpublished data). This, to our knowledge, is the first report of a pediocin-like molecule being produced by a member of the genus *Streptococcus*, and indicates that the production of molecules of this family may be more widespread than previously recognized.

4.3.2 Type IIb: Multi-Component Bacteriocins

Some non-lantibiotic bacteriocins, as for their lantibiotic counterparts, can require two or more peptides to effect optimal inhibitory activity. For detailed descriptions of the numerous two-component bacteriocins characterized, the reader is referred to the recent comprehensive review by Garneau et al. (2002). It has been proposed that two-component bacteriocins be subdivided into synergistic (S)- and enhancing (E)-type inhibitory agents

(Marciset et al. 1997). S-type two-component bacteriocin activities are dependent on the concerted action of both peptides, and neither component appears inhibitory on its own (Marciset et al. 1997; Garneau et al. 2002). Examples of S-type bacteriocin systems include lactococcin G from *L. lactis*, lactacin F from *Lactobacillus johnsonii*, and lactocin 705 from *Lactobacillus casei* (Nissen-Meyer et al. 1992; Allison et al. 1994; Cuozzo et al. 2000). Conversely, for an E-type two-component bacteriocin, either each component peptide or only one peptide of the duet possesses inhibitory activity, but combination of the components results in greatly enhanced killing action toward the target species. Thermophilin 13 from *Streptococcus thermophilus*, enterocin L50 from *Enterococcus faecium*, and ABP-118 from *Lactobacillus salivarius* are representatives of E-type two-component bacteriocins (Marciset et al. 1997; Cintas et al. 2000; Flynn et al. 2002).

Although very uncommon, reports of three- or four-component bacteriocin systems have also arisen (Donvito et al. 1997; Netz et al. 2001). For example, the SLUSH β-hemolysin produced by *Staphylococcus lugdunensis*, which also exhibits antimicrobial activity, apparently consists of three peptides (Donvito et al. 1997). Similarly, aureocin A70 elaborated by *St. aureus* is proposed to comprise four peptides (Netz et al. 2001). Whilst definitive classification of the SLUSH peptides as either an S- or an E-type system has not been possible, as each component peptide has not been individually purified, aureocin A70 can be regarded as an E-type system due to the intrinsic inhibitory activity of AucA, AucB and AucC, but not AucD (Netz et al. 2001).

4.3.3 Type IIc: Miscellaneous Unmodified Bacteriocins

Due to the absence of any constraints imposed by either physical structure or characteristics of their genetic loci, all single-peptide non-modified bacteriocins that do not fulfill the criteria of type IIa or type IIb are automatically members of type IIc (formerly class IId, according to the scheme of Cotter et al. 2005b). Type IIc is a menagerie of inhibitory agents produced by strains from many ecological sources. As a result, its members are by far the most diverse, for example, with regard to the post-translational processing of the prebacteriocins and export of the biologically active agents. A recent intriguing example of a type IIc bacteriocin is sakacin Q, in which the prebacteriocin and immunity-related protein are translationally coupled (Mathiesen et al. 2005). The identification of new type IIc bacteriocins has been further facilitated by the availability – for data-mining – of a multitude of complete genome sequences of Gram-positive bacteria. For example, bioinformatics and mutational analyses were used to detect type IIc mutacins produced by *S. mutans* UA159, the genome sequence reference strain (Hale et al. 2005b).

Many type IIc bacteriocins share common features such as a prepeptide possessing a GG-containing leader sequence, an associated three-domain ABC transporter, and an immunity-related protein. However, some type IIc members are exported, apparently via ABC transport systems, but without a recognizable N-terminal signal peptide. Examples include enterocin Q (from *E. faecium*), aureocin A53 (from *St. aureus*), and BHT-B (from *S. rattus*; Cintas et al. 2000; Netz et al. 2002; Hyink et al. 2005). It should also be noted that bacteriocin systems categorized as type IIb, such as the two-component enterocin L50 and the four-component aureocin A70 (Cintas et al. 2000; Netz et al. 2001), are also exported without identifiable signal peptides, indicating that this particular mode of export may be a common phenomenon among Gram-positive bacteria.

In the following subsections, we wish to digress from the bacteriocins of LAB by (1) describing the unusual type IIc bacteriocins from the propionic acid bacteria, and (2) highlighting the roles of type IIc-like non-inhibitory peptides in the so-called three-component signal transduction systems.

4.3.3.1 The Propionic Acid Bacteria: Producers of Novel Bacteriocins

While most of the bacteriocins described in this chapter originate from LAB, several novel bacteriocins produced by members of the propionic acid bacteria are of special interest (Table 4.2). These bacteria produce propionic acid as the primary end-product of glucose fermentation. The prominent bacteriocinogenic species belong to the genus *Propionibacterium*, principally those involved in the production of Swiss-style cheeses (Faye et al. 2000; Brede et al. 2004). Despite numerous reports of bacteriocinogenicity by this group of bacteria (Faye et al. 2002), only five propionibacterial bacteriocins have been characterized to date, four of which are non-lantibiotic peptides (Table 4.2) and the fifth a large, ca. 20-kDa protein (Miescher et al. 2000). The first to be described biochemically was propionicin PLG-1, a ca. 10-kDa bacteriocin produced by *Propionibacterium thoenii* P127 (Lyon and Glatz 1993), a strain that elaborates two propionicin activities (Ben-Shushan et al. 2003). However, neither the N-terminal amino acid sequence nor the genetic determinant of PLG-1 have been reported.

By contrast, the second inhibitory agent elaborated by *P. thoenii* P127, designated GBZ-1 (Ben-Shushan et al. 2003), has been characterized at the genetic level and is highly homologous to a novel protease-activated antimicrobial protein (PAMP) produced by *Propionibacterium jensenii* (Faye et al. 2002). Based on sequence analysis of its structural gene, it has been hypothesized that PAMP is initially synthesized as a 225-aa precursor protein containing a 27-residue Sec-dependent signal peptide (Faye et al. 2002). Upon export, an additional processing step involving proteolytic cleavage (at a specific Arg–Arg site) by an as yet unidentified extracellular protease yields the mature, biologically active 64-aa PAMP molecule (Faye et al. 2002). It is interesting to note that the post-export processing of closticin 574, a non-lantibiotic bacteriocin produced by *Clostridium tyrobutyricum* (Kemperman et al. 2003a), is analogous to that of PAMP. In addition to PAMP, certain strains of

Table 4.2 Some distinctive class II bacteriocins and their characteristics

Bacteriocin	Producer species	Distinctive characteristics	Key reference(s)
Type IIa: pediocin-like peptides			
Pediocin PA1	*Pediococcus acidilactici*	Prototype anti-listerial bacteriocin; biosynthetic locus is plasmid-borne; contains YGNGV motif that typifies this bacteriocin family; mode of action and mechanism of immunity are well-defined; first bacteriocin shown to be exported by an ABC transporter with concomitant cleavage of a signal peptide immediately following a double-glycine (GG) sequence motif	Rodriguez et al. (2002), Fimland et al. (2005)
Carnobacteriocin B2	*Carnobacterium piscicola*	First pediocin-like bacteriocin shown to be exported via a Sec-dependent transport system	Rodriguez et al. (2002), Fimland et al. (2005)
Listeriocin 743A	*Listeria innocua*	First bacteriocin to be described for *Listeria* spp.; displays the typical characteristics of pediocin-like peptides and is exported via the Sec-dependent route	Kalmokoff et al. (2001)
Ubericin A	*Streptococcus uberis*	First streptococcal pediocin-like inhibitory agent to be characterized; chromosomally encoded biosynthetic locus	G.A. Burtenshaw, unpublished data
Type IIb: multi-component inhibitors			
Lactococcin G	*Lactococcus lactis*	First two-component and S (synergistic)-type class II bacteriocin to be characterized	Nissen-Meyer et al. (1992)
Thermophilin 13	*Streptococcus thermophilus*	Prototype E (enhancer)-type two-component bacteriocin; ThmA* is the primary inhibitory peptide and ThmB* (no intrinsic activity of its own), when added, enhances the inhibitory activity; export mechanism unknown, although GG motif is present in the putative leader sequences of the precursor peptides	Marciset et al. (1997)

(*Continued*)

Table 4.2 Some distinctive class II bacteriocins and their characteristics—Continued

Bacteriocin	Producer species	Distinctive characteristics	Key reference(s)
Enterocin L50	*Enterococcus faecium*	An E-type bacteriocin system in which the component peptides are secreted without recognizable N-terminal signal peptides	Cintas et al. (1998, 2000)
Streptocins STH$_1$ and STH$_2$	*Streptococcus gordonii*	A novel two-component bacteriocin/hemolysin system of which the biosynthesis is dependent upon development of natural competence for genetic transformation; expression of the structural genes appears to be activated by the competence-specific sigma factor, and the prepeptides are processed and exported via the same transport system required for secretion of the competence-stimulating peptide	N.C.K. Heng et al., submitted
Plantaricin NC8	*Lactobacillus plantarum*	A two-component bacteriocin of which the biogenesis is activated by a three-component signal transduction system that appears to be responsive to interspecies cell-to-cell contact	Maldonado et al. (2004)
SLUSH peptides	*Staphylococcus lugdunensis*	A putative three-component bacteriocin/hemolysin system; the individual peptides have not been purified	Donvito et al. (1997)
Aureocin A70	*Staphylococcus aureus*	A putative E-type four-component bacteriocin, but inhibitory activity can be demonstrated only with three individual component peptides; each component peptide appears to be secreted without a signal peptide	Netz et al. (2001)
Plantaricin C11	*Lb. plantarum*	A multi-component bacteriocin system encoded by two operons; genetic regulation of the bacteriocin locus is unique, in that two transcriptional regulators with antagonistic functions appear to operate, promoting environment-dependent temporal and spatial fine-tuning	Diep et al. (2003)

Type IIc: miscellaneous bacteriocins

Aureocin A53	*Staphylococcus aureus*	A single-peptide bacteriocin that, similarly to aureocin A70, is exported despite not possessing an identifiable signal peptide	Netz et al. (2002)
Sakacin Q	*Lactobacillus sakei*	A novel bacteriocin in which the precursor peptide and its immunity-related protein are translationally coupled	Mathiesen et al. (2005)
Mutacins IV and V	*Streptococcus mutans*	Identified by a combination of protein purification, bioinformatic analyses and genetic dissection methods; structural genes are located in different segments of the genome but both bacteriocins are exported by a single ABC transport system	Qi et al. (2001), Hale et al. (2005a, 2005b)
Sakacins T and X	*Lb. sakei*	Two bacteriocin systems encoded by divergent genetic loci that are activated by a single three-component regulatory system; the component peptides of both bacteriocins are exported by the same ABC transporter without the need for an accessory factor	Vaughan et al. (2003)

Bacteriocins produced by propionic acid bacteria (PAB)

Propionicin PLG-1	*Propionibacterium thoenii*	First bacteriocin from PAB to be characterized biochemically (but not genetically), with an estimated molecular mass of ca. 10 kDa	Lyon and Glatz (1993)
Propionicin T1 (thoenicin 447)	*P. thoenii*	Sec-dependent bacteriocin active against *Propionibacterium acnes* (causative agent of acne vulgaris)	Faye et al. (2000), van der Merwe et al. (2004)
Protease-activated antimicrobial protein (PAMP)/propionicin GBZ-1	*Propionibacterium jensenii* and *P. thoenii*	A novel bacteriocin that is exported (as a 198-aa precursor protein) via the Sec-dependent pathway, but is only activated when an additional 134 aa is cleaved off by extracellular proteases	Faye et al. (2002), Ben-Shushan et al. (2003)
Propionicin F	*Propionibacterium freudenreichii*	A novel bacteriocin, in that the biologically active peptide is the central portion of a much larger preprotein that is doubly processed	Brede et al. (2004)

P. thoenii have also been shown to produce propionicin T1 (thoeniicin 447), a Sec-exported non-lantibiotic peptide that is active against the acne-causing bacterium *Propionibacterium acnes* (Faye et al. 2000; van der Merwe et al. 2004).

One of the more unique bacteriocins encountered to date, and the most extensively studied of the propionicins, is propionicin F produced by *Propionibacterium freudenreichii* (Brede et al. 2004). Propionicin F constitutes the central 43-aa peptide portion, namely, residues Trp^{102} to Pro^{145}, of a much larger 255-aa precursor protein, PcfA (Brede et al. 2004). As the propionicin F genetic locus has been completely sequenced, the proteolytic processing steps (between Cys^{101} and Trp^{102}, and between Pro^{145} and Gly^{146}) are considered to be mediated by the gene products of *pcfB* and *pcfC*, respectively (Brede et al. 2004). Since PcfA does not contain an identifiable N-terminal signal peptide, the mechanism of export of propionicin F remains unknown, although PcfD, a putative ABC transporter (which lacks the peptidase domain), has been implicated here (Brede et al. 2004). However, while not explicitly stated by Brede et al. (2004), we have noticed that the amino acid sequence of mature propionicin F does contain a GG motif. Based on the analogy that the pediocin PA-1 prepeptide is 80% active (Sect. 4.3.1), it is therefore feasible that despite being exported by PcfD with an intact GG motif, propionicin F peptide itself is inherently biologically active. However, it is tempting to speculate whether a further processed propionicin F (i.e., minus the putative GG-containing signal peptide) may have higher levels of antimicrobial activity.

4.3.3.2 Bacteriocin-like Peptides as Signaling Molecules

As described in Section 4.2.1, bacteriocins such as nisin can influence expression of their biosynthetic operons through binding to their cognate two-component histidine kinase-response regulator signal transduction systems, which in turn regulates transcription of the bacteriocin operon. In recent years, it has become apparent that such "three-component regulatory mechanisms" are also involved in cell density-dependent phenomena (quorum sensing) and non-lantibiotic bacteriocin biosynthesis (see reviews by Kleerebezem and Quadri 2001; Morrison 2002). Activation of these processes is usually mediated by the binding of a specific inducing peptide to its cognate sensor histidine kinase, which in turn phosphorylates a dedicated response regulator. The latter then proceeds to up- or down-regulate the expression of genes under its control. Interestingly, the inducing peptides, which do not appear to exhibit intrinsic inhibitory activity, are synthesized as prepeptides containing signal peptides with double-glycine motifs. Secretion of these prepeptides into the extracellular milieu is invariably facilitated by dedicated ABC transporters not unlike those involved in the export of pediocin-like peptides (Sect. 4.3.1).

A classic biological phenomenon in Gram-positive bacteria that involves quorum sensing and a three-component signal transduction system is the development of natural competence for genetic transformation (reviewed by Morrison 2002; Lacks 2004). Competence is defined as the transient physiological state that allows a bacterium to take up DNA from the environment. Transformation of the recipient cell occurs when the exogenous DNA is integrated into the genome by homologous recombination and subsequently expressed. In the archetypal, naturally transformable Gram-positive bacterium, *Streptococcus pneumoniae*, the competence cascade is initially activated by binding of a 17-mer peptide, called the competence-stimulating peptide (CSP), to its cognate histidine kinase ComD, followed by phosphorylation of the response regulator ComE (Morrison 2002; Lacks 2004). Phosphorylated ComE perpetuates the signal by up-regulating expression of the *comCDE* and *comAB* operons as well as activating the expression of ComX, the alternative sigma factor involved in the expression of the genes encoding the DNA uptake machinery. The ComAB operon encodes the ABC transporter responsible for the export of CSP.

In our laboratory, we have recently shown that the production of two bacteriocins (STH_1 and STH_2) by the naturally competent oral bacterium *Streptococcus gordonii* is not only dependent upon activation of the competence cascade, but is also under the control of ComR, the homologue of the *S. pneumoniae* competence-specific sigma factor ComX (N.C.K. Heng, J.R. Tagg, G.R. Tompkins, submitted). Moreover, the bacteriocin precursor peptides, SthA and SthB, each contain a GG cleavage motif, and are exported via the ComAB (CSP-secreting) ABC transporter. Competence-associated lantibiotic (e.g., Smb; see Sect. 4.2.4) and non-lantibiotic (e.g., mutacin IV) bacteriocin production has also been shown to occur in *S. mutans* (van der Ploeg 2005; Yonezawa and Kuramitsu 2005; Kreth et al. 2005). However, in contrast to *S. gordonii*, (1) the promoters of the bacteriocin-encoding genes appear to contain nucleotide motifs for DNA-binding response regulators, possibly ComE (van der Ploeg 2005; Yonezawa and Kuramitsu 2005), and (2) export of the non-lantibiotic bacteriocins is mediated by an ABC transporter that differs from that involved in CSP secretion (Hale et al. 2005a).

In *Lactobacillus plantarum*, peptide-activated expression of bacteriocin (plantaricin) production has been very well characterized. For example, the complex plantaricin C11 biosynthetic locus comprises five distinct operons (*plnABCD*, *plnEFI*, *plnJKLR*, *plnMNOP* and *plnGHSTUV*), the first of which (*plnABCD*) encodes a unique three-component signal transduction system, but with two response regulators PlnC and PlnD (Diep et al. 2003). The *plnEFI* and *plnJKLR* operons encode peptides exhibiting bona fide inhibitory activity (and their associated immunity proteins), and *plnGHSTUV* specifies the components of the bacteriocin export machinery. The function of the last operon *plnMNOP* is unknown. It is believed that both PlnC and PlnD are antagonistic in their regulatory activity, which facilitates temporal and spatial fine-tuning of plantaricin production according to the environmental situation (Diep et al. 2003). A recent exciting development, demanding further investigation, has been the discovery of plantaricin biosynthesis (in *Lb. plantarum* strain NC8), which is activated by a three-component signal transduction system that appears to be responsive to interspecies cell–cell contact (Maldonado et al. 2004).

Bacteriocin (sakacin) biosynthesis mediated by three-component signal transduction systems has also been extensively characterized in various strains of the food-associated species *Lactobacillus sakei* (Vaughan et al. 2003), and although their mechanisms of genetic regulation are not as elaborate as that of plantaricin C11, they are no less intriguing. Notable examples include (1) sakacin A biosynthesis, which has been shown to be profoundly affected at 34–37 °C but not at 30–33 °C, due to temperature-dependent synthesis of the inducer peptide Sap-Ph (Diep et al. 2000), and (2) activation, by a single three-component system (encoded by *stxPRK*), of two adjacent bacteriocin systems (the two-component and single-peptide sakacins T and X, respectively) that are encoded by divergent operons (Vaughan et al. 2003, 2004). Moreover, in the latter case, the inducing peptide and the sakacins are exported by the same ABC transporter that, unlike that of many such secretion systems, does not appear to require an accessory factor (Vaughan et al. 2003).

4.4 Class III: The Large (>10 kDa) Bacteriocins

While the bacteriocins characterized from Gram-positive species are predominantly small (<10 kDa) peptides, several large antimicrobial proteins have been described at both the biochemical and genetic level. These bacteriocins typically manifest as heat-labile proteins, but one apparent exception is propionicin SM1, a heat-stable inhibitory agent produced by *P. jensenii* (Miescher et al. 2000). It should be noted, however, that aggregates of small peptides, for example, staphylococcin 1580 (Sahl 1994), have caused confusion in the past with regard to estimation of protein size. The bona fide large bacteriocins of Gram-positive bacteria can generally be subdivided into two distinct groups: (1) the bacteriolytic enzymes (or bacteriolysins), which facilitate the killing of sensitive strains by cell lysis, and (2) the non-lytic antimicrobial proteins. However, in some cases, such as propionicin SM1 and albusin B (from *Ruminococcus albus*; Chen et al. 2004), the lack of mode of action data precludes them at present from placement in this classification scheme (cf. Table 4.3).

4.4.1 Type IIIa: The Bacteriolysins (Bacteriolytic Enzymes)

4.4.1.1 Lysostaphin – The Prototype Bacteriolysin

Originally described more than 40 years ago (Schindler and Schuhardt 1964), lysostaphin (produced by *Staphylococcus simulans* biovar *staphylolyticus*)

Table 4.3 The Class III (>10 kDa) bacteriocins of Gram-positive bacteria

Bacteriocin	Producer species	Distinctive features	Relevant reference(s)
Bacteriolytic enzymes (bacteriolysins)			
Lysostaphin	*Staphylococcus simulans* subsp. *staphylolyticus*	Prototype and best-characterized bacteriolytic enzyme; plasmid-encoded; comprises two distinct domains separated by a linker; post-translational processing events, mode of action and mechanism of immunity well-defined	Schindler and Schuhardt (1964), King et al. (1980), DeHart et al. (1995), Thumm and Gotz (1997), Ehlert et al. (2000)
Zoocin A	*Streptococcus equi* subsp. *zooepidemicus*	First chromosomally encoded (and streptococcal) bacteriolysin to be characterized; attacks the interpeptide crosslink of streptococcal peptidoglycan; similarly to lysostaphin, is composed of two domains; function of zoocin immunity factor (Zif) similar to lysostaphin counterpart (Lif/Epr); biosynthesis of zoocin A may be subject to catabolite repression	Simmonds et al. (1996, 1997), Beatson et al. (1998), Lai et al. (2002), O'Rourke et al. (2003)
Millericin B	*Streptococcus milleri*	Novel broad-spectrum bacteriolysin that cleaves not only the interpeptide crosslinks but also the stem peptide of peptidoglycan	Beukes et al. (2000), Beukes and Hastings (2001)
Enterolysin A	*Enterococcus faecalis*	Relatively broad-spectrum bacteriolytic enzyme that may also cleave the stem peptide; C-terminal domain appears very similar to that of bacteriophage lysins	Nilsen et al. (2003)
Other large bacteriocins			
Helveticin J	*Lactobacillus helveticus*	Possibly the first chromosomally encoded and largest non-lytic bacteriocin (37 kDa) to be described; exact mode of action and mechanism of immunity unknown	Joerger and Klaenhammer (1986, 1990)
Dysgalacticin	*Streptococcus dysgalactiae* subsp. *equisimilis*	Plasmid-encoded non-lytic 21-kDa bacteriocin with a narrow spectrum of activity; possesses a novel predicted secondary structure containing a single essential disulfide bond	Wong et al. (1981), Heng et al. (2006)

(*Continued*)

Table 4.3 The Class III (>10 kDa) bacteriocins of Gram-positive bacteria—Continued

Bacteriocin	Producer species	Distinctive features	Relevant reference(s)
Streptococcin A-M57	*Streptococcus pyogenes* M-types 57 and 69	First plasmid-encoded large (17 kDa) streptococcal bacteriocin to be reported; exhibits an unusual inhibitory spectrum, but possesses similar predicted secondary structure to that of dysgalacticin	Simpson and Tagg (1984), Heng et al. (2004)
Propionicin SM1	*Propionibacterium jensenii*	20-kDa heat-stable bacteriocin identified in the propionic acid bacteria; mode of action and mechanism of immunity unknown	Miescher et al. (2000)
Albusin B	*Ruminococcus albus*	First large bacteriocin to be identified in a ruminal bacterium; mode of action and immunity mechanism unknown; structural gene encodes a prebacteriocin containing an unusually long secretion signal peptide	Chen et al. (2004)

represents the prototype bacteriolysin and is probably the most extensively studied large bacteriocin elaborated by any Gram-positive bacterium. Lysostaphin is a plasmid-encoded glycylglycine endopeptidase that kills sensitive cells by specifically hydrolyzing the pentaglycine crossbridges in peptidoglycan (Robinson et al. 1979; King et al. 1980). A homologue of lysostaphin, ALE-1 from *Staphylococcus capitis*, has also been characterized (Sugai et al. 1997a). Due to the ability of lysostaphin to kill members of virtually all staphylococcal species, including those that impact on human and animal health, such as *St. aureus* and *Staphylococcus epidermidis*, various reports over the years have recommended its use in a variety of medical and agricultural applications (Oldham and Daley 1991; Wu et al. 2003; Shah et al. 2004).

Lysostaphin is initially synthesized as a 493-aa precursor protein (preprolysostaphin), which comprises the 246-aa mature form of the bacteriocin plus the following N-terminal extensions: (1) a 36-aa secretion signal peptide at its N-terminus, followed by (2) 195 aa organized into 15 tandem repeats of a 13-aa sequence (Heinrich et al. 1987; Thumm and Gotz 1997). Following export (with concomitant removal of the signal peptide), the tandem repeats are removed by a cysteine protease to yield the fully activated lysostaphin molecule (Neumann et al. 1993; Thumm and Gotz 1997). The lysostaphin molecule is predicted to consist of two distinct domains separated by a short linker sequence: (1) a N-terminal peptidase domain responsible for the catalytic activity of the protein, and (2) a C-terminal targeting domain involved in binding to the peptidoglycan substrate (Wang et al. 1991; Simmonds et al. 1997).

Further studies showed that the genetic determinant conferring immunity to lysostaphin was also located on the plasmid (Heath et al. 1989). The immunity factor, designated *lif* or *epr*, encodes a protein that displays homology to the FemAB complex responsible for adding glycine residues to the pentaglycine crosslinks (Heath et al. 1989; Thumm and Gotz 1997). However, Epr adds serine residues, rather than glycine, and this change in the amino acid composition of the crosslinks is sufficient to protect the cell from the lytic effects of lysostaphin (Robinson et al. 1979; DeHart et al. 1995; Thumm and Gotz 1997; Sugai et al. 1997b; Ehlert et al. 2000). Overall, the findings arising from the studies on lysostaphin and its immunity factor have provided invaluable knowledge not only to researchers working on bacteriocins but also to those trying to elucidate the complexities of cell wall construction.

4.4.1.2 Zoocin A and Other Bacteriolysins

Within the last 10 years, much progress has also been made in the characterization of bacteriolysins produced by lactic acid bacteria, mainly from members of the genera *Streptococcus* and *Enterococcus*. The prototype streptococcal bacteriolytic enzyme is zoocin A, which is specified by a chromosomally located gene (*zooA*) in *Streptococcus equi* subsp. *zooepidemicus*. Despite exhibiting limited amino acid sequence similarity, zoocin A and Zif (the zoocin A immunity factor) share common properties with lysostaphin and Epr, respectively, such as (1) the hydrolysis of streptococcal interpeptide crossbridges, (2) a modular structure consisting of an N-terminal M37-like peptidase domain and a C-terminal substrate-binding domain, and (3) Zif, similarly to Epr, resembles a FemAB-like protein that, when expressed in a heterologous host such as *S. gordonii* (a zoocin A-susceptible species), confers the expected zoocin-resistant phenotype (Simmonds et al. 1996, 1997; Beatson et al. 1998; Liang et al. 2004). Intriguingly, Zif does not appear to alter the glycine–serine ratios of the interpeptide chain (Beatson et al. 1998), and therefore the exact mechanism of immunity to zoocin A remains enigmatic. A more recent and exciting development is the novel observation that the biosynthesis of zoocin A may be influenced by glucose levels, i.e., it may be catabolite-repressed (O'Rourke et al. 2003). In our laboratory, we have recently identified stellalysin, a new zoocin A-like antimicrobial protein produced by the oral bacterium *Streptococcus constellatus* subsp. *constellatus*. Preliminary analyses indicate that stellalysin biosynthesis may also be catabolite-repressed (N.C.K. Heng et al., unpublished data).

Aside from zoocin A and stellalysin, only two other bacteriolytic enzymes produced by lactic acid bacteria have been described, namely, millericin B from *Streptococcus milleri* and enterolysin A from *E. faecalis* (Beukes et al. 2000; Nilsen et al. 2003). Millericin B is distinctive in its ability to hydrolyze the cell walls of species such as *M. luteus*, *Staph. aureus* and non-millericin B-producing strains of *S. milleri*, all of which possess different interpeptide crosslinks (Beukes et al. 2000). It was further shown that millericin B could cleave peptidoglycan either in the stem peptide (which is common to the above-listed three species) or in the interpeptide crosslinks (Beukes et al. 2000). Moreover, the mechanism of immunity to millericin B, similarly to that of lysostaphin, involves amino acid substitution (leucine for threonine) in the interpeptide crosslinks of peptidoglycan (Beukes and Hastings 2001).

Enterolysin A is the first large bacteriocin to be described from *E. faecalis*, and similarly to millericin B, exhibits a rather diverse inhibitory spectrum. Although the common element in the peptidoglycan of all enterolysin A-sensitive species appears to be the stem peptide (L-Ala-D-Glu-L-Lys-D-Ala; Nilsen et al. 2003), the exact mode of action of enterolysin A remains to be determined. Enterolysin A is composed of the two-domain structure typical of other bacteriolysins (Nilsen et al. 2003). Interestingly, while the N-terminal domain of enterolysin A, similarly to that of other bacteriolysins, is of the M37-like peptidase type, the C-terminal domain displays significant homology to the lysins of *Lactobacillus casei* bacteriophages (Nilsen et al. 2003).

4.4.2 Type IIIb: The Non-Lytic Bacteriocins

As the antithesis to the bacteriolysins, several large bacteriocins have been shown to kill target cells by non-lytic means. This could involve dissipation of the proton motive force, leading to ATP starvation and ultimately cell death.

The first non-lytic bacteriocin to be described at the biochemical and genetic level was helveticin J, a 37-kDa bacteriocin produced by *Lactobacillus helveticus* that primarily targets other *Lactobacillus* species (Joerger and Klaenhammer 1986, 1990). However, the precise mode of action of helveticin J remains unknown.

Dysgalacticin (21 kDa) and streptococcin A-M57 (SA-M57; 17 kDa) are secreted bacteriocins produced by *Streptococcus dysgalactiae* subsp. *equisimilis* and M-type 57 *S. pyogenes*, respectively (Wong et al. 1981; Heng et al. 2004, 2006). The inhibitory spectrum of dysgalacticin is fairly narrow and is limited to strains of Lancefield serogroups A, C and G (Wong et al. 1981; Tagg and Wong 1983). On the other hand, the range of organisms inhibited by SA-M57 is unusual, consisting mainly of non-streptococcal Gram-positive species including *M. luteus*, *L. lactis*, all tested species of *Listeria* (including *Lis. monocytogenes*), *Bacillus megaterium* and *St. simulans* (Simpson and Tagg 1983; Heng et al. 2004). Both bacteriocins appear to kill sensitive cells in a non-lytic fashion (Wong et al. 1981; Simpson and Tagg 1983; Heng et al. 2006), although the exact mechanism remains unclear.

At first glance, the similarities between dysgalacticin and SA-M57 appear superficial (Heng et al. 2004, 2006): (1) the structural genes for both dysgalacticin (*dysA*) and SA-M57 (*scnM57*) are plasmid-borne, and (2) both bacteriocins are exported via Sec-dependent systems. Dysgalacticin does not display any similarity either to proteins of known function or to hypothetical proteins in publicly available databases. Conversely, SA-M57 exhibits primary amino acid sequence similarity with two hypothetical, potentially secreted proteins, EF1097 and YpkK, from *E. faecalis* and *Corynebacterium jeikeium*, respectively (Heng et al. 2004).

Despite the obvious lack of sequence similarity between dysgalacticin, SA-M57, EF1097 and YpkK, all four proteins possess similar predicted secondary structures consisting of (1) a fairly unstructured N-terminal portion, (2) a C-terminal region that appears to contain a helix-loop-helix motif, and (3) two cysteine residues that are predicted to form a disulfide bond. We have subsequently shown, for both dysgalacticin and SA-M57, that the two cysteines do indeed form a disulfide bond essential for antimicrobial activity (N.C.K. Heng et al., unpublished data). Furthermore, we have successfully expressed the EF1097 and YpkK structural genes in *E. coli*, and found that both recombinant proteins exhibit antimicrobial activity, with the former displaying a much broader inhibitory spectrum (P.M. Swe and H.J. Baird, unpublished data). Taken collectively, dysgalacticin, SA-M57, EF1097 and YpkK potentially constitute a novel family of antimicrobial proteins.

4.5 Class IV: The Cyclic Bacteriocins

Based on our proposed classification scheme for antimicrobial proteins produced by Gram-positive bacteria, the fourth and arguably the most unique class of bacteriocins is that encompassing the cyclic bacteriocins (Table 4.4).

Table 4.4 Known circular (cyclic) bacteriocins produced by Gram-positive bacteria

Bacteriocin	Producer species	Distinguishing features	Key reference(s)
Enterocin AS-48	*Enterococcus faecalis*	Prototype and most well-characterized circular bacteriocin, both biochemically and genetically	Maqueda et al. (2004)
Gassericin A	*Lactobacillus gasseri*	The first cyclic bacteriocin to be reported that contains a mixture of D- and L-amino acids (at least D-Ala)	Kawai et al. (2004b)
Reutericin 6	*Lactobacillus reuteri*	Identical primary amino acid sequence to gassericin A but different inhibitory spectrum due to altered composition of D-alanine	Kawai et al. (2004b)
Circularin A	*Clostridium beijerinckii*	Well-characterized cyclic bacteriocin with unusual post-translational processing; genetic locus (*cir*) contains genes homologous to those found in the AS-48 locus	Kemperman et al. (2003a, 2003b)
Butyrivibriocin A	*Butyrivibrio fibrisolvens*	The first cyclic bacteriocin to be identified from a rumen bacterium; similar to gassericin A	Kalmokoff et al. (2003)
Uberolysin	*Streptococcus uberis*	First cyclic bacteriocin to be characterized (biochemically and genetically) from the genus *Streptococcus*; unusual due to its thermolability and lysis of actively growing cells	R.E. Wirawan et al., submitted

These circular inhibitory agents are ribosomally synthesized peptides, which are post-translationally processed such that the first and last amino acids of the mature peptide are covalently bonded, corresponding to the so-called head-to-tail ligation (Maqueda et al. 2004). To date, this class comprises only a handful of members, the prototype being enterocin AS-48 (extensively reviewed by Maqueda et al. 2004).

4.5.1 Enterocin AS-48

More than 20 years ago, a new heat-stable inhibitory agent (designated enterocin AS-48) produced by *E. faecalis* subsp. *liquefaciens* strain S-48 was first described (Galvez et al. 1985). The broad inhibitory spectrum of enterocin AS-48 includes Gram-positive as well as certain Gram-negative species (Maqueda et al. 2004). Due to an inconsistency of nomenclature, AS-48 and its natural variants are also known by other names such as enterocin 4 and bacteriocin 21 (Maqueda et al. 2004). The cyclic nature of AS-48 caused initial attempts to obtain the amino acid sequence of the peptide to fail, since the N-terminus is essentially blocked. This limitation was eventually overcome by endopeptidase digestion followed by N-terminal sequencing of the internal fragments (Maqueda et al. 2004). AS-48 has since become one of the most extensively characterized bacteriocins in terms of its biochemistry (including a three-dimensional structure) and genetics.

Enterocin AS-48 is plasmid-encoded and its biosynthetic locus contains ten genes (Martinez-Bueno et al. 1998; Maqueda et al. 2004). The AS-48 structural gene, *as-48A*, specifies a 105-aa prepeptide containing a 35-residue leader peptide. Whereas *as-48BC$_1$D* are believed to be responsible for the maturation and secretion of AS-48, *as-48D$_1$EFGH* have been assigned roles in producer self-protection or immunity (Diaz et al. 2003). Upon cleavage of the leader peptide (by an as yet unidentified process), the Met^1 and Trp^{70} residues of the 70-aa linear form of AS-48 are then covalently bonded, with the concomitant loss of a water molecule (Maqueda et al. 2004). The three-dimensional NMR structure of AS-48 reveals that it has five alpha-helices that fold into a very compact structure, with the head-to-tail union (between Met^1 and Trp^{70}) residing in helix 5 (Maqueda et al. 2004). The latter is perceived to contribute to the heat stability and structural integrity of the molecule (Maqueda et al. 2004). It is noteworthy that the secondary structure of enterocin AS-48 resembles that of a mammalian lysin found in natural killer cells, which is a non-circular protein composed of five helices stabilized by disulfide bridges (González et al. 2000).

More recently, the biochemical and genetic characteristics of circularin A, a new cyclic bacteriocin produced by *Clostridium beijerinckii*, have been reported (Kemperman et al. 2003a, 2003b). The *cir* locus shares some elements with the *as-48* locus, albeit with limited similarity at the amino acid level. For example, *cirBCDE* are the putative counterparts of *as-48BCDD$_1$* (Kemperman et al. 2003b). Circularin A itself possesses an unusual leader peptide of only three amino acids, and exhibits limited amino acid similarity to AS-48 (Kawai et al. 2004b).

4.5.2 Gassericin A and Reutericin 6

Gassericin A and reutericin 6 are cyclic peptides produced by *Lactobacillus gasseri* and *Lactobacillus reuteri*, respectively, which possess identical primary amino acid sequences deduced from their structural genes (Kawai et al. 2004a). However, their inhibitory spectra as well as their killing kinetics against selected indicator bacteria differ (Kawai et al. 2004a). The basis for these phenotypic differences was revealed by partial composition analysis of the D- and L-amino acids of both peptides. It was determined that gassericin A and reutericin 6 both contain D- and L-amino acids, a novel finding in itself, but differing in their D-Ala:L-Ala ratios (Kawai et al. 2004a). It is not known whether other differences exist between the two peptides, as the evaluation of D- or L-status was carried out with only five of the 17 amino acids known to be present in both bacteriocins (Kawai et al. 2004a). Two other bacteriocins, butyrivibriocin AR10 (from *Butyrivibrio fibrisolvens*) and acidocin B (from *Lactobacillus acidophilus*), display significant amino acid similarity to gassericin A (Leer et al. 1995; Kalmokoff et al. 2003). Whereas butyrivibriocin AR10 has been shown to be circular, the physical properties of acidocin B remain to be confirmed. Interestingly, the structural gene of acidocin B (*acdB*; Leer et al. 1995) was reported prior to that of gassericin A (Kawai et al. 1998). It should be noted that of the four cyclic peptides described in this section, genetic data beyond that of the structural gene encoding each peptide exist only for butyrivibriocin AR10 (Kalmokoff et al. 2003). A detailed comparative analysis of the maturation- and transport-associated genes of these bacteriocins would greatly aid our understanding of their biogenesis.

4.5.3 Uberolysin

S. uberis, one of the causative agents of bovine mastitis, is a prolific producer of bacteriocins. Individual strains produce different combinations of various classes of bacteriocins, including the lantibiotic nisin U (Sect. 4.2.1), the pediocin-like ubericin A (Sect. 4.3.1), and a novel circular bacteriocin called uberolysin (R.E. Wirawan et al., submitted). In contrast to cyclic peptides such as AS-48, uberolysin is a ca. 7-kDa heat-labile cyclic bacteriocin that lyses only actively growing cells. The uberolysin biosynthetic locus, designated *ubl*, has been completely sequenced and contains five genes, two of which (*ublB* and *ublD*) display limited homology to the putative maturation (*cirB*) and transport (*cirD*) genes of the circularin A locus, respectively

(Kemperman et al. 2003b). The 76-aa uberolysin precursor peptide is deduced to possess an atypical 6-aa leader peptide, and circularization is predicted to occur between Leu^{+1} and Trp^{+70}. It is envisaged that a cocktail of various *S. uberis*-derived bacteriocins could be developed as an effective preventative agent against bovine mastitis.

4.6 Concluding Remarks

And so as researchers, advantaged by increasingly sophisticated genetic and protein technologies, continue to delve both deeper and more expansively within the amazing repertoire of antimicrobials invented by microbes to advantage their own survival, we can eagerly anticipate further big surprises. It is important to appreciate that bacteriocins, defined as such for our own convenience, represent only one facet of a probably seamless continuum of bacterial antimicrobial activities. By necessity, the modified scheme for the classification of bacteriocins of Gram-positive bacteria that we have proposed here will inevitably continue to evolve. We have attempted to present the reader with some views (albeit colored by our own experiences) of a knowledge slice from the now vast, but still mounting, literature in this field. It is difficult to predict what might be the highlights or even the trends of a review of this field a decade from now. Nevertheless, we are tempted to anticipate more interest in the larger (class III) bacteriocins and further knowledge of the factors, especially of the molecular mechanisms, influencing expression of bacteriocins in natural ecosystems. Perhaps we may also see further developments and successes in the Pasteurian application of bacteriocin-producing bacteria to infection control – the great man would have been well-pleased by that!

Acknowledgments. The financial support of the Health Research Council of New Zealand, the Otago Medical Research Foundation, University of Otago Research Grants Committee and the University of Otago Oral Microbiology and Dental Health Research Theme is gratefully acknowledged. Much gratitude is also extended to our colleagues who have provided pre-publication material for this chapter.

References

Allison GE, Fremaux C, Klaenhammer TR (1994) Expansion of bacteriocin activity and host range upon complementation of two peptides encoded within the lactacin F operon. J Bacteriol 176:2235–2241

Aranha C, Gupta S, Reddy KV (2004) Contraceptive efficacy of antimicrobial peptide nisin: in vitro and in vivo studies. Contraception 69:333–338

Beatson SA, Sloan GL, Simmonds RS (1998) Zoocin A immunity factor: a *femA*-like gene found in a group C streptococcus. FEMS Microbiol Lett 163:73–77

Ben-Shushan G, Zakin V, Gollop N (2003) Two different propionicins produced by *Propionibacterium thoenii* P-127. Peptides 24:1733–1740

Beukes M, Hastings JW (2001) Self-protection against cell wall hydrolysis in *Streptococcus milleri* NMSCC 061 and analysis of the millericin B operon. Appl Environ Microbiol 67:3888–3896

Beukes M, Bierbaum G, Sahl HG, Hastings JW (2000) Purification and partial characterization of a murein hydrolase, millericin B, produced by *Streptococcus milleri* NMSCC 061. Appl Environ Microbiol 66:23–28

Bibel DJ, Aly R, Bayles C, Strauss WG, Shinefield HR, Maibach HI (1983) Competitive adherence as a mechanism of bacterial interference. Can J Microbiol 29:700–703

Booth MC, Bogie CP, Sahl HG, Siezen RJ, Hatter KL, Gilmore MS (1996) Structural analysis and proteolytic activation of *Enterococcus faecalis* cytolysin, a novel lantibiotic. Mol Microbiol 21:1175–1184

Brede DA, Faye T, Johnsborg O, Ødegård I, Nes IF, Holo H (2004) Molecular and genetic characterization of propionicin F, a bacteriocin *Propionibacterium freudenreichii*. Appl Environ Microbiol 70:7303–7310

Breukink E, Wiedemann I, van Kraaij C, Kuipers OP, Sahl H, de Kruijff B (1999) Use of the cell wall precursor lipid II by a pore-forming peptide antibiotic. Science 286:2361–2364

Brotz H, Bierbaum G, Markus A, Molitor E, Sahl HG (1995) Mode of action of the lantibiotic mersacidin: inhibition of peptidoglycan biosynthesis via a novel mechanism? Antimicrob Agents Chemother 39:714–719

Brotz H, Bierbaum G, Reynolds PE, Sahl HG (1997) The lantibiotic mersacidin inhibits peptidoglycan biosynthesis at the level of transglycosylation. Eur J Biochem 246:193–199

Brotz H, Josten M, Wiedemann I, Schneider U, Gotz F, Bierbaum G, Sahl HG (1998) Role of lipid-bound peptidoglycan precursors in the formation of pores by nisin, epidermin and other lantibiotics. Mol Microbiol 30:317–327

Chatterjee S, Chatterjee S, Lad SJ, Phansalkar MS, Rupp RH, Ganguli BN, Fehlhaber HW, Kogler H (1992) Mersacidin, a new antibiotic from *Bacillus*. Fermentation, isolation, purification and chemical characterization. J Antibiot (Tokyo) 45:832–838

Chatterjee C, Paul M, Xie L, van der Donk WA (2005) Biosynthesis and mode of action of lantibiotics. Chem Rev 105:633–684

Chen J, Stevenson DM, Weimer PJ (2004) Albusin B, a bacteriocin from the ruminal bacterium *Ruminococcus albus* 7 that inhibits growth of *Ruminococcus flavefaciens*. Appl Environ Microbiol 70:3167–3170

Chikindas ML, Novak J, Driessen AJ, Konings WN, Schilling KM, Caufield PW (1995) Mutacin II, a bactericidal antibiotic from *Streptococcus mutans*. Antimicrob Agents Chemother 9:2656–2660

Cintas LM, Casaus P, Havarstein LS, Hernandez PE, Nes IF (1997) Biochemical and genetic characterization of enterocin P, a novel *sec*-dependent bacteriocin from *Enterococcus faecium* P13 with a broad antimicrobial spectrum. Appl Environ Microbiol 63:4321–4330

Cintas LM, Casaus P, Holo H, Hernandez PE, Nes IF, Havarstein LS (1998) Enterocins L50A and L50B, two novel bacteriocins from *Enterococcus faecium* L50, are related to staphylococcal hemolysins. J Bacteriol 180:1988–1994

Cintas LM, Casaus P, Herranz C, Havarstein LS, Holo H, Hernandez PE, Nes IF (2000) Biochemical and genetic evidence that *Enterococcus faecium* L50 produces enterocins L50A and L50B, the *sec*-dependent enterocin P, and a novel bacteriocin secreted without an N-terminal extension termed enterocin Q. J Bacteriol 182:6806–6814

Coburn PS, Gilmore MS (2003) The *Enterococcus faecalis* cytolysin: a novel toxin active against eukaryotic and prokaryotic cells. Cell Microbiol 5:661–669

Coburn PS, Hancock LE, Booth MC, Gilmore MS (1999) A novel means of self-protection, unrelated to toxin activation, confers immunity to the bactericidal effects of the *Enterococcus faecalis* cytolysin. Infect Immun 67:3339–3347

Coburn PS, Pillar CM, Jett BD, Haas W, Gilmore MS (2004) *Enterococcus faecalis* senses target cells and in response expresses cytolysin. Science 306:2270–2272

Cotter PD, Hill C, Ross RP (2005a) Bacterial lantibiotics: strategies to improve therapeutic potential. Curr Prot Pept Sci 6:61–75

Cotter PD, Hill C, Ross RP (2005b) Bacteriocins: developing innate immunity for food. Nature Rev Microbiol 3:777–788

Cuozzo SA, Sesma F, Palacios JM, de Ruiz Holgado AP, Raya RR (2000) Identification and nucleotide sequence of genes involved in the synthesis of lactocin 705, a two-peptide bacteriocin from *Lactobacillus casei* CRL 705. FEMS Microbiol Lett 185:157–161

Day AM, Cove JH, Phillips-Jones MK (2003) Cytolysin gene expression in *Enterococcus faecalis* is regulated in response to aerobiosis conditions. Mol Gen Genomics 269:31–39

de Vos WM, Kuipers OP, van der Meer JR, Siezen RJ (1995) Maturation pathway of nisin and other lantibiotics: post-translationally modified antimicrobial peptides exported by gram-positive bacteria. Mol Microbiol 17:427–437

DeHart HP, Heath HE, Heath LS, LeBlanc PA, Sloan GL (1995) The lysostaphin endopeptidase resistance gene (*epr*) specifies modification of peptidoglycan cross bridges in *Staphylococcus simulans* and *Staphylococcus aureus*. Appl Environ Microbiol 61:1475–1479

Diaz M, Valdivia E, Martinez-Bueno M, Fernandez M, Soler-Gonzalez AS, Ramirez-Rodrigo H, Maqueda M (2003) Characterization of a new operon, *as-48EFGH*, from the *as-48* gene cluster involved in immunity to enterocin AS-48. Appl Environ Microbiol 69:1229–1236

Diep DB, Axelsson L, Grefsli C, Nes IF (2000) The synthesis of the bacteriocin sakacin A is a temperature-sensitive process regulated by a pheromone peptide through a three-component regulatory system. Microbiology 146:2155–2160

Diep DB, Myhre R, Johnsborg O, Aakra A, Nes IF (2003) Inducible bacteriocin production in *Lactobacillus* is regulated by differential expression of the *pln* operons and by two antagonizing response regulators, the activity of which is enhanced upon phosphorylation. Mol Microbiol 47:483–494

Donvito B, Etienne J, Denoroy L, Greenland T, Benito Y, Vandenesch F (1997) Synergistic hemolytic activity of *Staphylococcus lugdunensis* is mediated by three peptides encoded by a non-*agr* genetic locus. Infect Immun 65:95–100

Ehlert K, Tschierske M, Mori C, Schroder W, Berger-Bachi B (2000) Site-specific serine incorporation by Lif and Epr into positions 3 and 5 of the staphylococcal peptidoglycan interpeptide bridge. J Bacteriol 182:2635–2638

Eijsink VG, Skeie M, Middelhoven PH, Brurberg MB, Nes IF (1998) Comparative studies of class IIa bacteriocins of lactic acid bacteria. Appl Environ Microbiol 64:3275–3281

Eijsink VG, Axelsson L, Diep DB, Havarstein LS, Holo H, Nes IF (2002) Production of class II bacteriocins by lactic acid bacteria; an example of biological warfare and communication. Antonie Van Leeuwenhoek 81:639–654

Faye T, Langsrud T, Nes IF, Holo H (2000) Biochemical and genetic characterization of propionicin T1, a new bacteriocin from *Propionibacterium thoenii*. Appl Environ Microbiol 66:4230–4236

Faye T, Brede DA, Langsrud T, Nes IF, Holo H (2002) An antimicrobial peptide is produced by extracellular processing of a protein from *Propionibacterium jensenii*. J Bacteriol 184:3649–3656

Fimland G, Blingsmo OR, Sletten K, Jung G, Nes IF, Nissen-Meyer J (1996) New biologically active hybrid bacteriocins constructed by combining regions from various pediocin-like bacteriocins: the C-terminal region is important for determining specificity. Appl Environ Microbiol 62:3313–3318

Fimland G, Johnsen L, Dalhus B, Nissen-Meyer J (2005) Pediocin-like antimicrobial peptides (class IIa bacteriocins) and their immunity proteins: biosynthesis, structure, and mode of action. J Pept Sci 11:688–696

Flynn S, van Sinderen D, Thornton GM, Holo H, Nes IF, Collins JK (2002) Characterization of the genetic locus responsible for the production of ABP-118, a novel bacteriocin produced by the probiotic bacterium *Lactobacillus salivarius* subsp. *salivarius* UCC118. Microbiology 148:973–984

Fredericq P (1946) Sur la pluralité des récepteurs d'antibiose de *E. coli*. C R Soc Biol 140:1189–1190

Galvez A, Valdivia E, Maqueda M, Montoya E (1985) Production of bacteriocin-like substances by group D streptococci of human origin. Microbios 43(176S):223–232

Garneau S, Martin NI, Vederas JC (2002) Two-peptide bacteriocins produced by lactic acid bacteria. Biochimie 84:577–592

Georgalaki MD, Van Den Berghe E, Kritikos D, Devreese B, Van Beeumen J, Kalantzopoulos G, De Vuyst L, Tsakalidou E (2002) Macedocin, a food-grade lantibiotic produced by *Streptococcus macedonicus* ACA-DC 198. Appl Environ Microbiol 68:5891–5903

Gilmore MS, Skaugen M, Nes I (1996) *Enterococcus faecalis* cytolysin and lactocin S of *Lactobacillus sake*. Antonie Van Leeuwenhoek 69:129–138

González C, Langdon GM, Bruix M, Gálvez A, Valdivia E, Maqueda M, Rico M (2000) Bacteriocin AS-48, a microbial cyclic polypeptide structurally and functionally related to mammalian NK-lysin. Proc Natl Acad Sci USA 97:11221–11226

Gratia A (1925) Sur un remarquable exemple d'antagonisme entre deux souches de colibacille. C R Soc Biol 93:1040–1042

Gravesen A, Ramnath M, Rechinger KB, Andersen N, Jansch L, Hechard Y, Hastings JW, Knochel S (2002) High-level resistance to class IIa bacteriocins is associated with one general mechanism in *Listeria monocytogenes*. Microbiology 148:2361–2369

Gross E, Morell JL (1971) The structure of nisin. J Am Chem Soc 93:4634–4635

Guder A, Schmitter T, Wiedemann I, Sahl HG, Bierbaum G (2002) Role of the single regulator MrsR1 and the two-component system MrsR2/K2 in the regulation of mersacidin production and immunity. Appl Environ Microbiol 68:106–113

Haas W, Shepard BD, Gilmore MS (2002) Two-component regulator of *Enterococcus faecalis* cytolysin responds to quorum-sensing autoinduction. Nature 415:84–87

Hale JDF, Heng NCK, Jack RW, Tagg JR (2005a) Identification of *nlmTE*, the locus encoding the ABC transport system required for export of nonlantibiotic mutacins in *Streptococcus mutans*. J Bacteriol 187:5036–5039

Hale JDF, Ting Y-T, Jack RW, Tagg JR, Heng NCK (2005b) Bacteriocin (mutacin) production by *Streptococcus mutans* genome sequence reference strain UA159: elucidation of the antimicrobial repertoire by genetic dissection. Appl Environ Microbiol 71:7613–7617

Hancock REW (1997) Peptide antibiotics. Lancet 349:418–422

Hastings JW, Sailer M, Johnson K, Roy KL, Vederas JC, Stiles ME (1991) Characterization of leucocin A-UAL 187 and cloning of the bacteriocin gene from *Leuconostoc gelidum*. J Bacteriol 173:7491–7500

Håvarstein LS, Holo H, Nes IF (1994) The leader peptide of colicin V shares consensus sequences with leader peptides that are common among peptide bacteriocins produced by gram-positive bacteria. Microbiology 140:2383–2389

Heath HE, Heath LS, Nitterauer JD, Rose KE, Sloan GL (1989) Plasmid-encoded lysostaphin endopeptidase resistance of *Staphylococcus simulans* biovar *staphylolyticus*. Biochem Biophys Res Commun 160:1106–1109

Heinrich P, Rosenstein R, Bohmer M, Sonner P, Gotz F (1987) The molecular organization of the lysostaphin gene and its sequences repeated in tandem. Mol Gen Genet 209:563–569

Heng NCK, Burtenshaw GA, Jack RW, Tagg JR (2004) Sequence analysis of pDN571, a plasmid encoding novel bacteriocin production in M-type 57 *Streptococcus pyogenes*. Plasmid 52:225–229

Heng NCK, Ragland NL, Swe PM, Inglis MA, Baird HJ, Tagg JR, Jack RW (2006) Dysgalacticin: a novel, plasmid-encoded bacteriocin produced by *Streptococcus dysgalactiae* subsp. *equisimilis*. Microbiology 152:1991–2001

Hille M, Kies S, Gotz F, Peschel A (2001) Dual role of GdmH in producer immunity and secretion of the staphylococcal lantibiotics gallidermin and epidermin. Appl Environ Microbiol 67:1380–1383

Hindre T, Le Pennec JP, Haras D, Dufour A (2004) Regulation of lantibiotic lacticin 481 production at the transcriptional level by acid pH. FEMS Microbiol Lett 231:291–298

Hyink O, Balakrishnan M, Tagg JR (2005) *Streptococcus rattus* strain BHT produces both a class I two-component lantibiotic and a class II bacteriocin. FEMS Microbiol Lett 252:235–241

Hynes WL, Tagg JR (1986) Proteinase-related broad-spectrum inhibitory activity among group-A streptococci. J Med Microbiol 22:257–264

Jack RW, Tagg JR (1991) Isolation and partial structure of streptococcin A-FF22. In: Jung G, Sahl H-G (eds) Nisin and novel lantibiotics. ESCOM, Leiden, pp 171–179

Jack RW, Tagg JR (1992) Factors affecting production of the group A streptococcus bacteriocin SA-FF22. J Med Microbiol 36:132–138

Jack RW, Carne A, Metzger J, Stefanovic S, Sahl HG, Jung G, Tagg J (1994) Elucidation of the structure of SA-FF22, a lanthionine-containing antibacterial peptide produced by *Streptococcus pyogenes* strain FF22. Eur J Biochem 220:455–462

Jack RW, Tagg JR, Ray B (1995) Bacteriocins of gram-positive bacteria. Microbiol Rev 59:171–200

Joerger MC, Klaenhammer TR (1986) Characterization and purification of helveticin J and evidence for a chromosomally determined bacteriocin produced by *Lactobacillus helveticus* 481. J Bacteriol 167:439–446

Joerger MC, Klaenhammer TR (1990) Cloning, expression and nucleotide sequence of the *Lactobacillus helveticus* 481 gene encoding the bacteriocin helveticin J. J Bacteriol 172:6339–6347

Jung G (1991) Lantibiotics: a survey. In: Jung G, Sahl H-G (eds) Nisin and novel lantibiotics. ESCOM, Leiden, pp 1–34

Jung G, Sahl H-G (1991) (eds) Nisin and novel lantibiotics. ESCOM, Leiden

Kalmokoff ML, Banerjee SK, Cyr T, Hefford MA, Gleeson T (2001) Identification of a new plasmid-encoded *sec*-dependent bacteriocin produced by *Listeria innocua* 743. Appl Environ Microbiol 67:4041–4047

Kalmokoff ML, Cyr TD, Hefford MA, Whitford MF, Teather RM (2003) Butyrivibriocin AR10, a new cyclic bacteriocin produced by the ruminal anaerobe *Butyrivibrio fibrisolvens* AR10: characterization of the gene and peptide. Can J Microbiol 49:763–773

Kanatani K, Oshimura M, Sano K (1995) Isolation and characterization of acidocin A and cloning of the bacteriocin gene from *Lactobacillus acidophilus*. Appl Environ Microbiol 61:1061–1067

Karaya K, Shimizu T, Taketo A (2001) New gene cluster for lantibiotic streptin possibly involved in streptolysin S formation. J Biochem (Tokyo) 129:769–775

Kawai Y, Saito T, Suzuki M, Itoh T (1998) Sequence analysis by cloning of the structural gene of gassericin A, a hydrophobic bacteriocin produced by *Lactobacillus gasseri* LA39. Biosci Biotechnol Biochem 62:887–892

Kawai Y, Ishii Y, Arakawa K, Uemura K, Saitoh B, Nishimura J, Kitazawa H, Yamazaki Y, Tateno Y, Itoh T, Saito T (2004a) Structural and functional differences in two cyclic bacteriocins with the same sequences produced by lactobacilli. Appl Environ Microbiol 70:2906–2911

Kawai Y, Kemperman R, Kok J, Saito T (2004b) The circular bacteriocins gassericin A and circularin A. Curr Prot Pept Sci 5:393–398

Kemperman R, Jonker M, Nauta A, Kuipers OP, Kok J (2003a) Functional analysis of the gene cluster involved in production of the bacteriocin circularin A by *Clostridium beijerinckii* ATCC 25752. Appl Environ Microbiol 69:5839–5848

Kemperman R, Kuipers A, Karsens H, Nauta A, Kuipers O, Kok J (2003b) Identification and characterization of two novel clostridial bacteriocins, circularin A and closticin 574. Appl Environ Microbiol 69:1589–1597

Kessler H, Steuernagel D, Gillessen D, Kamiyama T (1987) Complete sequence determination and localisation of one imino and three sulfide bridges of the nonadecapeptide Ro09-0198 by homonuclear 2D-NMR spectroscopy. The DQF-RELAYED_NOESY experiment. Helv Chim Acta 70:726–741

King BF, Biel ML, Wilkinson BJ (1980) Facile penetration of the *Staphylococcus aureus* capsule by lysostaphin. Infect Immun 29:892–896

Klaenhammer TR (1993) Genetics of bacteriocins produced by lactic acid bacteria. FEMS Microbiol Rev 12:39–85

Kleerebezem M, Quadri LE (2001) Peptide pheromone-dependent regulation of antimicrobial peptide production in Gram-positive bacteria: a case of multicellular behavior. Peptides 22:1579–1596

Koponen O, Takala TM, Saarela U, Qiao M, Saris PE (2004) Distribution of the NisI immunity protein and enhancement of nisin activity by the lipid-free NisI. FEMS Microbiol Lett 231:85–90

Kreth J, Merritt J, Shi W, Qi F (2005) Co-ordinated bacteriocin production and competence development: a possible mechanism for taking up DNA from neighbouring species. Mol Microbiol 57:392–404

Kruszewska D, Sahl HG, Bierbaum G, Pag U, Hynes SO, Ljungh A (2004) Mersacidin eradicates methicillin-resistant *Staphylococcus aureus* (MRSA) in a mouse rhinitis model. J Antimicrob Chemother 54:648–653

Kuipers OP, Beerthuyzen MM, de Ruyter PG, Luesink EJ, de Vos WM (1995) Autoregulation of nisin biosynthesis in *Lactococcus lactis* by signal transduction. J Biol Chem 270 45:27299–304

Lacks SA (2004) Transformation. In: Tuomanen EI, Mitchell TJ, Morrison DA, Spratt BG (eds) The pneumococcus. ASM Press, Washington, DC, pp 89–115

Lai AC, Tran S, Simmonds RS (2002) Functional characterization of domains found within a lytic enzyme produced by *Streptococcus equi* subsp. *zooepidemicus*. FEMS Microbiol Lett 215:133–138

Leer RJ, van der Vossen JM, van Giezen M, van Noort JM, Pouwels PH (1995) Genetic analysis of acidocin B, a novel bacteriocin produced by *Lactobacillus acidophilus* Microbiology 141:1629–1635

Liang Q, Simmonds RS, Timkovich R (2004) NMR evidence for independent domain structures in zoocin A, an antibacterial exoenzyme. Biochem Biophys Res Commun 317:527–530

Lyon WJ, Glatz BA (1993) Isolation and purification of propionicin PLG-1, a bacteriocin produced by a strain of *Propionibacterium thoenii*. Appl Environ Microbiol 59:83–88

Maldonado A, Ruiz-Barba JL, Jimenez-Diaz R (2004) Induction of plantaricin production in *Lactobacillus plantarum* NC8 after coculture with specific gram-positive bacteria is mediated by an autoinduction mechanism. J Bacteriol 186:1556–1564

Maqueda M, Gálvez A, Bueno MM, Sanchez-Barrena MJ, González C, Albert A, Rico M, Valdivia E (2004) Peptide AS-48: prototype of a new class of cyclic bacteriocins. Curr Prot Pept Sci 5:399–416

Marciset O, Jeronimus-Stratingh MC, Mollet B, Poolman B (1997) Thermophilin 13, a nontypical antilisterial poration complex bacteriocin, that functions without a receptor. J Biol Chem 272:14277–14284

Marki F, Hanni E, Fredenhagen A, Oostrum J (1991) Mode of action of the lanthionine-containing peptide antibiotics duramycin, duramycin B and C, and cinnamycin as indirect inhibitors of phospholipase A2. Biochem Pharmacol 42:2027–2035

Martin NI, Sprules T, Carpenter MR, Cotter PD, Hill C, Ross RP, Vederas JC (2004) Structural characterization of lacticin 3147, a two-peptide lantibiotic with synergistic activity. Biochemistry 43:3049–3056

Martinez-Bueno M, Valdivia E, Galvez A, Coyette J, Maqueda M (1998) Analysis of the gene cluster involved in production and immunity of the peptide antibiotic AS-48 in *Enterococcus faecalis*. Mol Microbiol 27:347–358

Mathiesen G, Huehne K, Kroeckel L, Axelsson L, Eijsink VG (2005) Characterization of a new bacteriocin operon in sakacin P-producing *Lactobacillus sakei*, showing strong translational coupling between the bacteriocin and immunity genes. Appl Environ Microbiol 71:3565–3574

Mattick ATR, Hirsch A (1947) Further observations on an inhibitory substance (nisin) produced by group N streptococci. Lancet ii:5–7

McAuliffe O, Hill C, Ross RP (2000) Each peptide of the two-component lantibiotic lacticin 3147 requires a separate modification enzyme for activity. Microbiology 146:2147–2154

McAuliffe O, O'Keeffe T, Hill C, Ross RP (2001a) Regulation of immunity to the two-component lantibiotic, lacticin 3147, by the transcriptional repressor LtnR. Mol Microbiol 39:982–993

McAuliffe O, Ross RP, Hill C (2001b) Lantibiotics: structure, biosynthesis and mode of action. FEMS Microbiol Rev 25:285–308

Miescher S, Stierli MP, Teuber M, Meile L (2000) Propionicin SM1, a bacteriocin from *Propionibacterium jensenii* DF1: isolation and characterization of the protein and its gene. Syst Appl Microbiol 23:174–184

Miller KW, Schamber R, Osmanagaoglu O, Ray B (1998) Isolation and characterization of pediocin AcH chimeric protein mutants with altered bactericidal activity. Appl Environ Microbiol 64:1997–2005

Morgan SM, O'Connor PM, Cotter PD, Ross RP, Hill C (2005) Sequential actions of the two component peptides of the lantibiotic lacticin 3147 explain its antimicrobial activity at nanomolar concentrations. Antimicrob Agents Chemother 49:2606–2611

Morrison DA (2002) Is anybody here? Cooperative bacterial gene regulation via peptide signals between Gram-positive bacteria. In: Hodgson DA, Thomas CM (eds) Signals, switches, regulons and cascades: control of bacterial gene expression. Cambridge University Press, Cambridge, pp 231–249

Mota-Meira M, LaPointe G, Lacroix C, Lavoie MC (2000) MICs of mutacin B-Ny266, nisin A, vancomycin, and oxacillin against bacterial pathogens. Antimicrob Agents Chemother 44:24–29

Naruse N, Tenmyo O, Tomita K, Konishi M, Miyaki T, Kawaguchi H, Fukase K, Wakamiya T, Shiba T (1989) Lanthiopeptin, a new peptide antibiotic. Production, isolation and properties of lanthiopeptin. J Antibiot (Tokyo) 42:837–845

Netz DJA, Sahl H-G, Marcelino R, Nascimento JS, de Oliveira SS, Soares MB, Bastos MCF (2001) Molecular characterisation of aureocin A70, a multi-peptide bacteriocin isolated from *Staphylococcus aureus*. J Mol Biol 311:939–949

Netz DJA, Pohl R, Beck-Sickinger AG, Selmer T, Pierik AJ, Bastos MCF, Sahl H-G (2002) Biochemical characterisation and genetic analysis of aureocin A53, a new, atypical bacteriocin from *Staphylococcus aureus*. J Mol Biol 319:745–756

Neumann VC, Heath HE, LeBlanc PA, Sloan GL (1993) Extracellular proteolytic activation of bacteriolytic peptidoglycan hydrolases of *Staphylococcus simulans* biovar *staphylolyticus*. FEMS Microbiol Lett 110:205–211

Nilsen T, Nes IF, Holo H (2003) Enterolysin A, a cell-wall-degrading bacteriocin from *Enterococcus faecalis* LMG 2333. Appl Environ Microbiol 69:2975–2984

Nissen-Meyer J, Holo H, Havarstein LS, Sletten K, Nes IF (1992) A novel lactococcal bacteriocin whose activity depends on the complementary action of two peptides. J Bacteriol 174:5686–5692

Oldham ER, Daley MJ (1991) Lysostaphin: use of a recombinant bactericidal enzyme as a mastitis therapeutic. J Dairy Sci 74:4175–4182

O'Rourke AD, Simmonds RS, Cook GM (2003) Catabolite repression of a bacteriocin-like inhibitory substance produced by *Streptococcus equi* subsp. *zooepidemicus*. In: Abstr Vol 103rd Annu Meet American Society for Microbiology, Washington, DC, Abstract H-006

Pag U, Heidrich C, Bierbaum G, Sahl H (1999) Molecular analysis of expression of the lantibiotic pep5 immunity phenotype. Appl Environ Microbiol 65:591–598

Paik SH, Chakicherla A, Hansen JN (1998) Identification and characterization of the structural and transporter genes for, and the chemical and biological properties of, sublancin 168, a novel lantibiotic produced by *Bacillus subtilis* 168. J Biol Chem 273:23134–23142

Palmer DE, Mierke DF, Pattaroni C, Goodman M, Wakamiya T, Fukase K, Kitazawa M, Fujita H, Shiba T (1989) Interactive NMR and computer simulation studies of lanthionine-ring structures. Biopolymers 28:397–408

Piard JC, Muriana PM, Desmazeaud MJ, Klaenhammer TR (1992) Purification and partial characterisation of lacticin 481, a lanthionine-containing bacteriocin produced by *Lactococcus lactis* subsp. *lactis* CNRZ 481. Appl Environ Microbiol 58:279–284

Qi F, Chen P, Caufield PW (1999) Purification of mutacin III from group III *Streptococcus mutans* UA787 and genetic analyses of mutacin III biosynthesis genes. Appl Environ Microbiol 65:3880–3887

Qi F, Chen P, Caufield PW (2000) Purification and biochemical characterization of mutacin I from the group I strain of *Streptococcus mutans*, CH43, and genetic analysis of mutacin I biosynthesis genes. Appl Environ Microbiol 66:3221–3229

Qi F, Chen P, Caufield PW (2001) The group I strain of *Streptococcus mutans*, UA140, produces both the lantibiotic mutacin I and a nonlantibiotic bacteriocin, mutacin IV. Appl Environ Microbiol 67:15–21

Qiao M, Immonen T, Koponen O, Saris PE (1995) The cellular location and effect on nisin immunity of the NisI protein from *Lactococcus lactis* N8 expressed in *Escherichia coli* and *L. lactis*. FEMS Microbiol Lett 131:75–80

Ramnath M, Beukes M, Tamura K, Hastings JW (2000) Absence of a putative mannose-specific phosphotransferase system enzyme IIAB component in a leucocin A-resistant strain of *Listeria monocytogenes*, as shown by two-dimensional sodium dodecyl sulfate-polyacrylamide gel electrophoresis. Appl Environ Microbiol 66:3098–3101

Rawlinson EL, Nes IF, Skaugen M (2002) LasX, a transcriptional regulator of the lactocin S biosynthetic genes in *Lactobacillus sakei* L45, acts both as an activator and a repressor. Biochimie 84:559–567

Reddy KV, Aranha C, Gupta SM, Yedery RD (2004) Evaluation of antimicrobial peptide nisin as a safe vaginal contraceptive agent in rabbits: in vitro and in vivo studies. Reproduction 128:117–126

Reis M, Eschbach-Bludau M, Iglesias-Wind MI, Kupke T, Sahl HG (1994) Producer immunity towards the lantibiotic Pep5: identification of the immunity gene *pepI* and localization and functional analysis of its gene product. Appl Environ Microbiol 60:2876–2883

Riley MA, Wertz JE (2002) Bacteriocins: evolution, ecology, and application. Annu Rev Microbiol 56:117–137

Rince A, Dufour A, Uguen P, Le Pennec JP, Haras D (1997) Characterization of the lacticin 481 operon: the *Lactococcus lactis* genes *lctF*, *lctE*, and *lctG* encode a putative ABC transporter involved in bacteriocin immunity. Appl Environ Microbiol 63:4252–4260

Rink R, Kuipers A, de Boef E, Leenhouts KJ, Driessen AJ, Moll GN, Kuipers OP (2005) Lantibiotic structures as guidelines for the design of peptides that can be modified by lantibiotic enzymes. Biochemistry 44:8873–8882

Robinson JM, Hardman JK, Sloan GL (1979) Relationship between lysostaphin endopeptidase production and cell wall composition in *Staphylococcus staphylolyticus*. J Bacteriol 137:1158–1164

Rodriguez JM, Martinez MI, Kok J (2002) Pediocin PA-1, a wide-spectrum bacteriocin from lactic acid bacteria. Crit Rev Food Sci Nutr 42:91–121

Rogers LA (1928) The inhibitory effect of *Streptococcus lactis* on *Lactobacillus bulgaricus*. J Bacteriol 16:321–325

Ross KF, Ronson CW, Tagg JR (1993) Isolation and characterization of the lantibiotic salivaricin A and its structural gene *salA* from *Streptococcus salivarius* 20P3. Appl Environ Microbiol 59:2014–2021

Ryan MP, Jack RW, Josten M, Sahl HG, Jung G, Ross RP, Hill C (1999) Extensive post-translational modification, including serine to D-alanine conversion, in the two-component lantibiotic, lacticin 3147. J Biol Chem 274:37544–37550

Sahl HG (1994) Staphylococcin 1580 is identical to the lantibiotic epidermin: implications for the nature of bacteriocins from gram-positive bacteria. Appl Environ Microbiol 60:752–755

Schindler CA, Schuhardt VT (1964) Lysostaphin: a new bacteriolytic agent for the staphylococcus. Proc Natl Acad Sci USA 51:414–421

Schlegel R, Slade HD (1973) Properties of a *Streptococcus sanguis* (group H) bacteriocin and its separation from the competence factor of transformation. J Bacteriol 115:655–661

Schneider TR, Karcher J, Pohl E, Lubini P, Sheldrick GM (2000) Ab initio structure determination of the lantibiotic mersacidin. Acta Crystallogr Sect D Biol Crystallogr 56:705–713

Schnell N, Entian KD, Schneider U, Gotz F, Zahner H, Kellner R, Jung G (1988) Prepeptide sequence of epidermin, a ribosomally synthesized antibiotic with four sulphide-rings. Nature 333:276–278

Shah A, Mond J, Walsh S (2004) Lysostaphin-coated catheters eradicate *Staphylococccus aureus* challenge and block surface colonization. Antimicrob Agents Chemother 48:2704–2707

Simmonds RS, Pearson L, Kennedy RC, Tagg JR (1996) Mode of action of a lysostaphin-like bacteriolytic agent produced by *Streptococcus zooepidemicus* 4881. Appl Environ Microbiol 62:4536–4541

Simmonds RS, Simpson WJ, Tagg JR (1997) Cloning and sequence analysis of *zooA*, a *Streptococcus zooepidemicus* gene encoding a bacteriocin-like inhibitory substance having a domain structure similar to that of lysostaphin. Gene 189:255–261

Simpson WJ, Tagg JR (1983) M-type 57 group A streptococcus bacteriocin. Can J Microbiol 29:1445–1451

Simpson WJ, Tagg JR (1984) Survey of the plasmid content of group A streptococci. FEMS Microbiol Lett 23:195–199

Simpson WJ, Ragland NL, Ronson CW, Tagg JR (1995) A lantibiotic gene family widely distributed in *Streptococcus salivarius* and *Streptococcus pyogenes*. Dev Biol Stand 85:639–643

Stein T, Entian KD (2002) Maturation of the lantibiotic subtilin: matrix-assisted laser desorption/ionization time-of-flight mass spectrometry to monitor precursors and their proteolytic processing in crude bacterial cultures. Rapid Commun Mass Spectrom 16:103–110

Sugai M, Fujiwara T, Akiyama T, Ohara M, Komatsuzawa H, Inoue S, Suginaka H (1997a) Purification and molecular characterization of glycylglycine endopeptidase produced by *Staphylococcus capitis* EPK1. J Bacteriol 179:1193–1202

Sugai M, Fujiwara T, Ohta K, Komatsuzawa H, Ohara M, Suginaka H (1997b) *epr*, which encodes glycylglycine endopeptidase resistance, is homologous to *femAB* and affects serine content of peptidoglycan cross bridges in *Staphylococcus capitis* and *Staphylococcus aureus*. J Bacteriol 179:4311–4318

Tagg JR (1992) Bacteriocins of Gram-positive bacteria: an opinion regarding their nature, nomenclature and numbers. In: James R, Lazdunski C, Pattus F (eds) Bacteriocins, microcins and lantibiotics. Springer, Berlin Heidelberg New York, NATO ASI Series vol H65, pp 33–35

Tagg JR, Bannister LV (1979) "Fingerprinting" beta-haemolytic streptococci by their production of and sensitivity to bacteriocine-like inhibitors. J Med Microbiol 12:397–411

Tagg JR, Dierksen KP (2003) Bacterial replacement therapy: adapting 'germ warfare' to infection prevention. Trends Biotechnol 21:217–223

Tagg JR, Wong HK (1983) Inhibitor production by group G streptococci of human and of animal origin. J Med Microbiol 16:409–415

Tagg JR, Dajani AS, Wannamaker LW (1976) Bacteriocins of gram-positive bacteria. Bacteriol Rev 40:722–756

Thumm G, Gotz F (1997) Studies on prolysostaphin processing and characterization of the lysostaphin immunity factor (Lif) of *Staphylococcus simulans* biovar *staphylolyticus*. Mol Microbiol 23:1251–1265

Tompkins GR, Peavey MA, Birchmeier KR, Tagg JR (1997) Bacteriocin production and sensitivity among coaggregating and noncoaggregating oral streptococci. Oral Microbiol Immunol 12:98–105

Upton M, Tagg JR, Wescombe P, Jenkinson HF (2001) Intra- and interspecies signaling between *Streptococcus salivarius* and *Streptococcus pyogenes* mediated by SalA and SalA1 lantibiotic peptides. J Bacteriol 183:3931–3938

Vadyvaloo V, Arous S, Gravesen A, Hechard Y, Chauhan-Haubrock R, Hastings JW, Rautenbach M (2004) Cell-surface alterations in class IIa bacteriocin-resistant *Listeria monocytogenes* strains. Microbiology 150:3025–3033

van den Hooven HW, Lagerwerf FM, Heerma W, Haverkamp J, Piard JC, Hilbers CW, Siezen RJ, Kuipers OP, Rollema HS (1996) The structure of the lantibiotic lacticin 481 produced by *Lactococcus lactis*: location of the thioether bridges. FEBS Lett 391:317–322

van der Merwe IR, Bauer R, Britz TJ, Dicks LM (2004) Characterization of thoeniicin 447, a bacteriocin isolated from *Propionibacterium thoenii* strain 447. Int J Food Microbiol 92:153–160

van der Ploeg JR (2005) Regulation of bacteriocin production in *Streptococcus mutans* by the quorum-sensing system required for development of genetic competence. J Bacteriol 187:3980–3989

Vaughan A, Eijsink VG, van Sinderen D (2003) Functional characterization of a composite bacteriocin locus from malt isolate *Lactobacillus sakei* 5. Appl Environ Microbiol 69:7194–7203

Vaughan A, O'Mahony J, Eijsink VG, O'Connell-Motherway M, van Sinderen D (2004) Transcriptional analysis of bacteriocin production by malt isolate *Lactobacillus sakei* 5. FEMS Microbiol Lett 235:377–384

Venema K, Kok J, Marugg JD, Toonen MY, Ledeboer AM, Venema G, Chikindas ML (1995) Functional analysis of the pediocin operon of *Pediococcus acidilactici* PAC1.0: PedB is the immunity protein and PedD is the precursor processing enzyme. Mol Microbiol 17:515–522

Wang X, Wilkinson BJ, Jayaswal RK (1991) Sequence analysis of a *Staphylococcus aureus* gene encoding a peptidoglycan hydrolase activity. Gene 102:105–109

Wescombe PA, Tagg JR (2003) Purification and characterization of streptin, a type A1 lantibiotic produced by *Streptococcus pyogenes*. Appl Environ Microbiol 69:2737–2747

Wescombe PA, Heng NCK, Jack RW, Tagg JR (2005) Bacteriocins associated with cytotoxicity for eukaryotic cells. In: Proft T (ed) Microbial toxins: molecular and cellular biology. Horizon Bioscience, Wymondham, pp 399–448

Wescombe PA, Burton JP, Cadieux PA, Klesse NA, Hyink O, Heng NCK, Chilcott CN, Reid G, Tagg JR (2006a) Megaplasmids encode differing combinations of lantibiotics in *Streptococcus salivarius*. Antonie van Leeuwenhoek (DOI: 10.2007/s10482-006-9081-y)

Wescombe PA, Upton M, Dierksen KP, Ragland NL, Sivabalan S, Wirawan RE, Inglis MA, Moore CJ, Walker GV, Chilcott CN, Jenkinson HF, Tagg JR (2006b) Production of the lantibiotic salivaricin A and its variants by oral streptococci and use of a specific induction assay to detect their presence in human saliva. Appl Environ Microbiol 72:1459–1466

Widdick DA, Dodd HM, Barraille P, White J, Stein TH, Chater KF, Gasson MJ, Bibb MJ (2003) Cloning and engineering of the cinnamycin biosynthetic gene cluster from *Streptomyces cinnamoneus cinnamoneus* DSM 40005. Proc Natl Acad Sci USA 100:4316–4321

Wirawan RE, Klesse NA, Jack RW, Tagg JR (2006) Molecular and genetic characterization of a novel nisin produced by *Streptococcus uberis*. Appl Environ Microbiol 72:1148–1156

Wong HK, Tagg JR, Hynes WL (1981) Bacteriocin-like inhibitors of group A streptococci produced by group F and group G streptococci. Proc Univ Otago Med School 59:105–106

Wu JA, Kusuma C, Mond JJ, Kokai-Kun JF (2003) Lysostaphin disrupts *Staphylococcus aureus* and *Staphylococcus epidermidis* biofilms on artificial surfaces. Antimicrob Agents Chemother 47:3407–3414

Yonezawa H, Kuramitsu HK (2005) Genetic analysis of a unique bacteriocin, Smb, produced by *Streptococcus mutans* GS5. Antimicrob Agents Chemother 49:541–548

5 Peptide and Protein Antibiotics from the Domain *Archaea*: Halocins and Sulfolobicins

RICHARD F. SHAND[1] AND KATHRYN J. LEYVA[2]

Summary

Production of peptide or protein antibiotics is a near-universal feature in all three domains of life. While bacteriocins and eucaryocins have been studied for decades, research in the field of archaeocins (halocins and sulfolobicins) is just emerging; most *Archaea* have yet to be screened for antibiotic production. To date, only seven halocins and one sulfolobicin have been partially or fully characterized, but antagonism studies suggest that there are hundreds of different halocins. Halocins are diverse in size (ranging from 3–35 kDa), thermal stability, and salt-dependence. Their activity spectra are typically "broad" with respect to killing other haloarchaea, and some microhalocins (small peptide halocins) have demonstrated cross-phylum inhibition. Currently, the mechanism of action is known only for halocin H6/H7, which inhibits the Na^+/H^+ antiporter in both haloarchaeal and mammalian cells. The potential biotechnological applications of other halocins will hinge on discovery of their mechanisms of action.

5.1 Introduction

In contrast to the wealth of studies for bacteriocins that began in 1925 (Gratia 1925) and have been chronicled in this volume, the characterization of peptide and protein antibiotics from organisms that inhabit the domain *Archaea* ("archaeocins") is only beginning (O'Connor and Shand 2002) – the first report of an archaeocin was published in 1982 (Rodriguez-Valera et al. 1982). The term "archaeocin" was coined to distinguish peptide and protein antibiotics produced by *Archaea* from those produced by members of the domain *Bacteria* (Price and Shand 2000). To refer to archaeocins as bacteriocins

---

[1]Department of Biological Sciences, Northern Arizona University, Flagstaff, AZ 86011, USA, e-mail: richard.shand@nau.edu
[2]Department of Microbiology, Arizona College of Osteopathic Medicine, Midwestern University, Glendale, AZ 85308, USA, e-mail: kleyva@midwestern.edu

perpetuates the confusion between these two domains of prokaryotic organisms: *Archaea* are no more closely related to *Bacteria* than are *Eucarya* (Woese et al. 1990). Having made this distinction, it is logical to include a chapter on archaeocins in a text devoted to bacteriocins, as the archaeocin field is just emerging and grouping the prokaryotic antimicrobial producers together makes sense. In addition, the term "halobacteria" was used early on as a collective term that encompassed all extremely halophilic members of the domain *Archaea* (i.e., members of the archaeal family *Halobacteriaceae*) and not, as one would assume, as a reference to halophilic members of the domain *Bacteria*. Subsequently, this terminology has been replaced by the term "haloarchaea", preventing further confusion. Continuing with the same nomenclature, peptide and protein antibiotics produced by members of the domain *Eucarya* are called "eucaryocins" (O'Connor and Shand 2002), with the first reports appearing in the early 1960s. Consequently, there is a plethora of information about these protein antibiotics as well (see http://www.bbcm.univ.trieste.it/ for an up-to-date list of 880+ eucaryocins).

To date, archaeocins have been characterized from only two phylogenetic groups: euryarchaeal extreme halophiles (haloarchaea) that produce "halocins" (O'Connor and Shand 2002), and the crenarchaeal genus *Sulfolobus*, an aerobic hyperthermophile that produces a "sulfolobicin" (Prangishvili et al. 2000). Although "production of halocin is a practically universal feature of archaeal halophilic rods" (Torreblanca et al. 1994), and based upon antagonism studies (Meseguer et al. 1986; Torreblanca et al. 1994), there appear to be hundreds of different halocins, only a handful of these have been characterized (see Table 5.1). Halocin protein sequences are unique, as they do not match anything in the protein sequence databases. Unfortunately, of the four haloarchaeal genomes that have been sequenced (*Halobacterium* sp. NRC-1: Ng et al. 2000; *Haloferax volcanii*: www.tigr.org/tdb; *Haloarcula marismortui*: Baliga et al. 2004; and *Natronomonas pharaonis*, an alkaliphilic haloarchaeon: Falb et al. 2005), none is a halocin producer (see Sect. 5.2.1). At the moment, halocin research must take place in the absence of a fully sequenced genome containing a halocin gene. Despite this limitation, all haloarchaea are aerobes and are easy to grow, with typical generation times between 1.5 and 3 h (Robinson et al. 2005). Detailed protocols for isolating microhalocins are also available (Shand 2006), as is a complete bibliography of the halocin literature (http://jan.ucc.nau.edu/~shand). What this field needs now are more scientists.

5.2 Halocins

5.2.1 The Ubiquity of Halocin Production

In 1992, J.R. Tagg posited that bacteriocin production would be a near-universal feature of bacteria, given a sufficient number of indicator strains

Table 5.1 Halocin characteristics (this table is reprinted with permission from Springer; O'Connor and Shand 2002)

Halocin	Producer (source)	Size	GenBank accession #	Thermal stability	Salt dependent	Activity spectrum[a]	Mechanism	References
A4	Haloarchaeon TuA4 (solar saltern, Tunisia)	7,435 Da	–	≥ 1 week at boiling[b]	No	Broad Sulfolobus spp.	ND	Haseltine et al. (2001), Duncan (2004)
C8	Halobacterium strain AS7092 (Great Chaidan Salt Lake, China)	~31.1 kDa (prepro-protein; ProC8), 7,427 Da (mature; HalC8)	AY310321	> 60 min at 100 °C	No	Broad	ND	Li et al. (2003), Sun et al. (2005)
G1	Halobacterium strain GRB (solar saltern, France)	ND[c]	–	ND	ND	Broad	ND	Soppa and Oesterhelt (1989)
H1	Haloferax mediterranei M2a (previously Xia3; solar saltern, Spain)	31 kDa	–	< 50 °C	Yes	Broad	Membrane permeability	Rodriguez-Valera et al. (1982), Meseguer et al. (1986), Platas (1995), Platas et al. (1996), Platas et al. (2002)
H2	Haloarchaeon Gla2.2[d] (solar saltern, Spain)	ND	–	ND	ND	Broad	ND	Rodriguez-Valera et al. (1982)
H3	Haloarchaeon Gaa 12 (solar saltern, Spain)	ND	–	ND	ND	Broad	ND	Rodriguez-Valera et al. (1982), Meseguer et al. (1986)

(Continued)

Table 5.1 Halocin characteristics (this table is reprinted with permission from Springer; O'Connor and Shand 2002)—Continued

Halocin	Producer (source)	Size	GenBank accession #	Thermal stability	Salt dependent	Activity spectrum[a]	Mechanism	References
H4	*Haloferax mediterranei* R4 (solar saltern, Spain)	39.6 kDa (preprotein), 34.9 kDa (mature)	U16389	< 60 °C	Partially[f]	Broad	Proton flux?	Rodriguez-Valera et al. (1982), Meseguer and Rodriguez-Valera (1985), Meseguer and Rodriguez-Valera (1986), Meseguer et al. (1986), Meseguer et al. (1995), Cheung et al. (1997), Shand et al. (1997), Perez (1999), Perez (2000)
H5	Haloarchaeon Ma 2.20[d] (solar saltern, Spain)	ND	–	ND	ND	Narrow	ND	Rodriguez-Valera et al. (1982)

(*Continued*)

Table 5.1 Halocin characteristics (this table is reprinted with permission from Springer; O'Connor and Shand 2002)—Continued

Halocin	Producer (source)	Size	GenBank accession #	Thermal stability	Salt dependent	Activity spectrum[a]	Mechanism	References
H6/H7	*Haloferax gibbonsii* Ma2.39[e] (solar saltern, Spain)	~3 kDa	–	≤ 90 °C	No	Broad	Na$^+$/H$^+$ antiporter inhibitor	Rodriguez-Valera et al. (1982), Meseguer et al. (1986), Torreblanca et al. (1989), Meseguer et al. (1995), Alberola et al. (1998)
R1	*Halobacterium* strain GN101 (solar saltern, Mexico)	3.8 kDa	–	60 °C	No	Broad *Sulfolobus* spp., *M. thermophila*	ND	Rdest and Sturm (1987), Shand et al. (1999), Haseltine et al. (2001), O'Connor (2002)
S8	Haloarchaeon S8a (Great Salt Lake, UT)	33.9 kDa (preproprotein), 3.58 kDa (mature)	AF276080	≥ 24 h at boiling[b]	No	Broad *Sulfolobus* spp.	ND	Shand et al. (1999), Price and Shand (2000), Haseltine et al. (2001)

[a]Activity spectrum refers to inhibition of haloarchaea, unless otherwise indicated. For a definition of "broad", see Sect. 5.2.3
[b]This study was done at 2,113 m (7,000 ft); water boils at 93 °C at this elevation
[c]ND: not determined
[d]Isolates Gla2.2 and Ma2.20 were labeled as GLA22 and MA220, respectively, in Rodriguez-Valera et al. (1982)
[e]Halocin H6 is produced by *Haloferax gibbonsii* Ma2.39 and was first reported as a 32-kDa protein, but is now known to be a microhalocin of about 3 kDa. *Hfx. gibbonsii* Ma2.39 is proprietary and should not be confused with a different halocin-producing strain, *Hfx. gibbonsii* Ma2.38 (ATCC 33595). Halocin H7 is halocin H6, but is produced by a halocin-overproducing mutant of *Hfx. gibbonsii* Ma2.39 called *Hfx. gibbonsii* Alicante SPH7
[f]See Sect. 5.2.4

(Tagg 1992). Similarly, the diversity of eucaryal organisms that produce antimicrobial peptides is vast, ranging from protozoans to plants to humans (O'Connor and Shand 2002). After conducting two non-overlapping antagonism studies (Meseguer et al. 1986: 79 isolates; Torreblanca et al. 1994: 68 isolates), Torreblanca et al. reached the same conclusion regarding halocins: "Production of halocin is a practically universal feature of archaeal halophilic rods" (Torreblanca et al. 1994). In all, of the 147 isolates screened, only three failed to show any inhibitory activity.

However, there are three issues that surround these two antagonism studies. First, only a single medium with one salt concentration (25% (w/v) marine salts) was used to grow all of the isolates; no attempt was made to use optimal NaCl concentration(s) for growth of any of the isolates. As mentioned in the studies, this resulted in wide variations in growth rates. Second, it is not clear that all activities were due to peptides or proteins. In the 1986 study, the 79 isolates were assigned to one of 15 groups based in part on their activity spectrum. However, only supernatants from "representatives" of these 15 groups were subjected to protease inactivation. The 1994 study does not state unequivocally that all cells or culture supernatants demonstrating inhibitory activities were treated with proteases, but they very well may have been. Third, the 1986 study indicated that *Haloferax volcanii* DS2 inhibited six of the 79 isolates, three of which were culture collection strains of the same genus: *Har. vallismortis*, *Har. marismortui* (previously *Hbt. marismortui*) and *Har. hispanica* (previously *Hbt. hispanicum*). However, in the Shand laboratory, *Hfx. volcanii* DS2 does not inhibit any of these three strains. In addition, scrutiny of the *Har. marismortui* genome and the *Nmn. pharaonis* genome does not reveal any obvious halocin sequences. Moreover, these two strains inhibited only a single member of the 79-member collection in the antagonism study (Meseguer et al. 1986). These differences may be due simply to differences between the various isolates in the strain collections.

5.2.2 The Role of Halocins in the Environment and the Inability to Detect Halocin Activity in Hypersaline Crystallizer Ponds

Given the ubiquity of halocin production described above, one might predict that aquatic hypersaline environments might be replete with halocin activity. To test this hypothesis, Kis-Papo and Oren (2000) sampled four different crystallizer ponds; two ponds were sampled only once whereas the other two were sampled repeatedly and at different times of the year. These ponds contained large numbers of prokaryotic microorganisms ($8.4 \times 10^6 - 7.2 \times 10^8$, by direct cell counts) dominated by haloarchaea. Using 12 haloarchaeal indicator strains representing five genera, cell-free brines showed no evidence of halocin activity, regardless of the pond, even when some of the brines were concentrated as much as 53.5-fold. From one pond in Eilat, Israel, a collection of 41 haloarchaea were isolated, 29 of which showed halocin activity against at least one of the 12 indicator strains, demonstrating that halocin producers were present in the pond. The authors concluded that "One possibility is that under field conditions no significant quantities of halocins are produced and that halocins are unimportant in interspecies competition in hypersaline lakes" (Kis-Papo and Oren 2000). They added that halocins might have been present in the brines, but they might have bound non-specifically to the filter membranes (although control experiments suggest that this was not an issue) or they may have been degraded by proteases present in the brines during transport.

Another possibility is that halocins are produced in crystallizer ponds, but as soon as they bind to a target (or even bind non-specifically to debris?), they become inactive. To determine how quickly halocin activity would disappear, preliminary "disappearing halocin activity" experiments involving mixing halocin-laden supernatants with halocin-sensitive cells in broth have been performed (O'Connor and Shand 2002). Samples were removed periodically and assayed for halocin activity. The activity of halocin A4 disappeared in less than a minute, halocin R1 activity was reduced in as little as 5 min with some activity remaining after 24 h, and halocin S8 activity did not diminish at all. It is unknown if these preliminary experiments were conducted under saturating concentrations of halocin-to-target.

The teleological explanation for prokaryotic antimicrobial production in the environment has been to reduce competition and/or lyse cells to acquire nutrients. For halocins, evidence supporting the latter part of this model has been found by Platas et al. (1996). A halocin producer (*Hfx. mediterranei* Ma2, formerly Xia3) was mixed with a non-producer (*Hbt. salinarum*) in the absence of any nutrients. The producer strain was able to grow, presumably through the release of cellular contents of the non-producer.

However, the concept that antimicrobial production reduces competition (and therefore diversity) is being challenged; the presence of antimicrobials in the environment is thought to maintain or even increase species diversity through a rock-paper-scissors model (Lenski and Riley 2002; Kirkup and Riley 2004; see Chap. 6, this volume). In this model, which organism(s) dominates may change over time, and although some organisms may become rare, they nevertheless persist and do not disappear. For example, in the early phylogenetic placement of the haloarchaea, three organisms isolated from a solar saltern in Alicante, Spain (*Hfx. mediterranei*, *Hfx. gibbonsii* and *Har. hispanica*) helped to define three of the founding genera in the family *Halobacteriaceae*. Upon returning to the site a couple of years later, none of these organisms was recovered by culturing or by PCR (Rodríguez-Valera et al. 1999). Are these organisms gone, or have they simply become rare? If they have become rare, why is that? Furthermore, the presence of antimicrobials may generate sufficient selective pressure for spontaneous antimicrobial-resistant mutants to arise.

5.2.3 Activity Spectra

Activity spectra (or killing breadth) tend to be relatively narrow in bacteriocins, being limited to bacteria closely related to the producing strains (Riley and Wertz 2002). However, nisin, a lactococcal lantibiotic, inhibits the crenarchaeal hyperthermophile *Sulfolobus acidocaldarius* (P.D. Clark and D.W. Grogan, personal communication). This is the first example of *Bacteria/Archaea* cross-domain inhibition.

Table 5.1 includes the activity spectra of halocins that have been characterized either fully or partially. Initial reports describing a particular halocin frequently use a relatively small number of characterized haloarchaeal strains from culture collections to determine its activity spectrum (e.g., Rodriguez-Valera et al. 1982; Kis-Papo and Oren 2000; Li et al. 2003). Halocins H1, H2, H3, H4, H5 and H6/H7 have been retested against 79 haloarchaeal strains in the 1986 antagonism study. All six have a "broad" activity spectrum when it comes to inhibiting haloarchaeal isolates (inhibiting between 63-74; Meseguer et al. 1986). Together, activity spectra in the two antagonism studies ranged from strains that inhibited zero to those that inhibited as many as 74 strains.

In order to challenge gram-positive and gram-negative bacteria, halocins have to be desalted without losing activity, limiting the population of testable halocins to the microhalocins (see Sect. 5.2.4). No microhalocin has been shown to inhibit any bacterial organism. This is not an unexpected result, as the microhalocins that have been characterized have either little or no net charge and are unable to interact with the negatively charged bacterial membrane the way that many bacteriocins and eucaryocins do. Similarly, microhalocins do not inhibit lower eukaryotic microorganisms, including *Saccharomyces* spp. However, halocin H6/H7, which inhibits the haloarchaeal Na^+/H^+ antiporter (Meseguer et al. 1995), also inhibits the Na^+/H^+ antiporter in a dog model (Alberola et al. 1998). It is not known if halocin H6/H7 inhibits Na^+/H^+ antiporters in organisms evolutionarily intermediate between haloarchaea and mammals.

Halocins A4, R1 and S8 all inhibit *Sulfolobus* spp. (crenarchaeal hyperthermophiles that grow optimally at 80°C and pH 3), with halocin R1 also inhibiting *Methanosarcina thermophila* (a mesophilic methane-producing euryarchaeote; Haseltine et al. 2001; Table 5.1). Indeed, this is a broad spectrum of activity representing cross-phylum inhibition, as haloarchaea are in the phylum Euryarchaeota whereas *Sulfolobus* spp. are in the phylum Crenarchaeota. It may be that other halocins can inhibit distantly related archaeal organisms, but they have yet to be tested for this breadth of inhibition.

From a hypersaline field site in Utah, we have isolated more than 350 different extreme halophiles spanning all three domains, as determined by amplification of 16S or 18S rDNA sequences using domain-specific primers (P.J. Polsgrove, B.A. Roberts, M.A. Mishler, R.F. Shand, unpublished data). Preliminary antagonism studies employing 48 purified isolates show that 62% inhibited at least one of the other isolates, with some isolates inhibiting as many as 30 of the other 47 strains. This is consistent with the Kis-Papo and Oren study (2000) where 71% (29/41) of the isolates from the crystallizer pond in Eilat inhibited at least one member of the 12 tester strains. Despite the large number of antimicrobial producers at the Utah site, the microbial diversity appears very high, supporting the argument that antimicrobial production may contribute to the maintenance or enhancement of species diversity. Polsgrove et al. (P.J. Polsgrove, B.A. Roberts, M.A. Mishler, R.F. Shand, unpublished data) also found bacterial extreme halophiles are inhibiting haloarchaea, haloarchaea are inhibiting bacterial extreme halophiles, and there are several extremely halophilic fungi that inhibit both bacterial and archaeal isolates. This is the first time cross-domain inhibition has been shown to occur in the environment, and this site will serve as an excellent model to study environmental chemical warfare among the three domains.

5.2.4 Common Features of Halocins

Halocins are either peptide (≤ 10 kDa; "microhalocins") or protein (>10 kDa) antibiotics produced by members of the archaeal family *Halobacteriaceae*. With one exception, halocin genes are induced at the transition between exponential and stationary phases (halocin H1 is induced during exponential phase; Platas et al. 1996). All

- halocin genes are located on megaplasmids (aka "mini-chromosomes"),
- halocin genes have typical haloarchaeal TATA boxes and TFB recognition elements (BRE), although the TATA box element for halocin C8 is a bit closer to the start site of transcription than usual (18 bp rather than 22–25 bp; Sun et al. 2005),
- halocin transcripts are "leaderless", where the transcriptional start site is either coincident with or only a few bps upstream of the translational start codon ATG,
- halocin preproteins appear to be exported by the twin arginine translocation (Tat) pathway, as all have a Tat signal motif at their amino terminus,
- mature halocins are inactivated by one or more proteases, confirming their proteinaceous nature,
- microhalocins are hydrophobic and are robust, as they can be desalted without losing activity, are insensitive to organic solvents such as acetonitrile and acetone, are relatively insensitive to heat (halocin R1 is the most sensitive, but can withstand heating at 60 °C for 1 h without losing activity; Table 5.1; O'Connor 2002), and can be stored at 4 °C for prolonged periods (as long as 7 years for halocin R1; O'Connor 2002) without significant loss of activity, and
- protein halocins (halocins H1 and H4) are heat-labile and lose activity when desalted below 5% (w/v) NaCl, although halocin H4 can be desalted to 10 mM Na^+ with only a twofold loss in activity (Perez 2000; Table 5.1). However, desalting to this level decreases the length of time halocin H4 can be stored at 4 °C.

5.2.5 Microhalocins (≤ 10 kDa)

5.2.5.1 Halocin S8

Halocin S8 (HalS8), produced by the uncharacterized haloarchaeal strain S8a isolated from the Great Salt Lake, UT by Penny Amy, was the first microhalocin to be characterized at both the protein and genetic levels (Price and Shand 2000). The mature microhalocin is composed of 36 amino acids with a molecular mass of 3.58 kDa. HalS8 contains four cysteine residues, which may form two disulfide bridges. However, no information on the tertiary structure of this microhalocin (or any other halocin, for that matter) is available. Currently, there is no evidence that HalS8 undergoes any post-translational modification of the amino acid sequence, but this has yet to be verified. The *halS8* gene is composed of a 933-bp open reading frame, yielding a 311 amino acid preproprotein upon translation. Processing of the preproprotein yields three separate proteins or peptides: a 230 amino acid N-terminal protein containing a typical Tat signal sequence, a 45 amino acid C-terminal peptide, and in between, the 36 amino acid mature halocin. Liberation of the halocin from the interior of its preproprotein is unique. Whether its release is autocatalytic or due to a protease is unknown (see De Castro et al. 2006 for a review of haloarchaeal proteases). Price and Shand (2000) speculated that the 230 amino acid N-terminal protein and the 45 amino acid C-terminal peptide might play roles in halocin immunity, regulation, induction, and/or translocation. However, BLAST searches revealed no matches to any other sequence within the database that would help to elucidate their possible function(s).

5.2.5.2 Halocin R1

Halocin R1 (HalR1), the second microhalocin to be characterized, is produced by *Hbt. salinarum* GN101, originally isolated from a solar saltern in Guererro Negro, Mexico by Barbara Javor (Ebert et al. 1986). Initial studies found HalR1 to have a molecular mass of 6.2 kDa by SDS-PAGE (Rdest and Sturm 1987). In contrast, later experiments revealed that halocin R1 appeared to be attached to a "carrier" protein, giving an apparent mass of about 29 kDa by gel filtration during purification (Shand et al. 1999; O'Connor 2002). However, upon heating the halocin-laden material prior to gel filtration, the halocin dissociated from the "carrier" protein and eluted at its true mass of 3.8 kDa. The HalR1 peptide consists of 38 amino acids, as determined by Edman degradation, with striking similarity to HalS8: HalR1 is 63% identical (capitalized residues below) and 71% similar to HalS8 (Price and Shand 2000; O'Connor 2002; O'Connor and Shand 2002):

- HalR1: `lqsNINiNTAAaVILiFNQVqvgALCaPTpVsGGgPpP`
- HalS8: `sdcNINsNTAAdVILcFNQVgscALCsPTIV-GG-PvP`

The small, yet significant differences in the amino acid sequences of these two microhalocins must be responsible for the differences in activity spectra and thermolability (Price and Shand 2000; O'Connor 2002; O'Connor and Shand 2002), but exactly which residues are involved remains to be determined.

Rdest and Sturm (1987) demonstrated that HalR1 is archaeostatic, as no changes in optical density or cell morphology of sensitive *Hbt. salinarum* (formerly *Hbt. halobium*) cells were noted after incubation with HalR1 for 7 days, and the cultures were able to resume growth upon removal of the halocin. Additionally, no zones of inhibition were seen when HalR1 was spotted onto fully grown lawns of sensitive cells, demonstrating that HalR1 is not archaeolytic. The archaeostatic response is dose-dependent, as increasing amounts of HalR1 resulted in proportional increases in the degree of inhibition, determined by the effect on growth in broth (Rdest and Sturm 1987).

5.2.5.3 Halocin H6/H7

Halocin H7 (HalH7, formerly known as HalH6) is produced by *Hfx. gibbonsii* Ma2.39, originally isolated from a solar saltern near Alicante, Spain (Torreblanca et al. 1986). The molecular mass of halocin H7 was initially calculated to be 32 kDa by gel filtration (Torreblanca et al. 1989). Similar to halocin R1, denaturing conditions (in this case, SDS-PAGE) released the mature halocin from a larger "carrier" protein, yielding a peptide of approximately 3 kDa (I. Meseguer, personal communication). Therefore, this halocin is now reclassified as a microhalocin. Although the size of the protein has been elucidated, the gene and protein sequences unfortunately are proprietary. Stability studies have shown that HalH7 can be desalted and is heat-resistant, which is consistent with the physicochemical stability profile of the other microhalocins (see Sect. 5.2.4 and Table 5.1). Halocin H7 is archaeolytic, described as having "single-hit kinetics" (a linear, inverse relationship between survival of sensitive cells and halocin concentration; O'Connor and Shand 2002) in the range of 5–80 arbitrary units (AU)/ml (Torreblanca et al. 1989). Exposure of sensitive cells to HalH7 caused the cells to swell and eventually lyse, indicating that the target site of activity of HalH7 is the cell membrane (Torreblanca et al. 1989). Further studies examined the effect of HalH7 on changes in intracellular volume, internal pH, membrane potential, proton motive force, and ionic flux in sensitive cells; results showed that the specific target of HalH7 is the Na^+/H^+ antiporter (Meseguer et al. 1995). This is significant, as it not only provides the first specific mechanism of action that can be attributed to any halocin, but it has also been shown to inhibit both haloarchaeal and mammalian Na^+/H^+ antiporters (Meseguer et al. 1995; Alberola et al. 1998; see Sect. 5.3).

5.2.5.4 Halocin A4

Halocin A4 is produced by an uncharacterized haloarchaeon isolated from a Tunisian saltern by Felicitas Pfeifer. It has been purified from a concentrated culture supernatant, using gel filtration column chromatography and reversed-phase HPLC as described in Shand (2006). The molecular mass of halocin A4 is 7,435 Da, as determined by mass spectrometry (very similar to halocin C8; see Sect. 5.2.5.5), and it is both acidic (pI = 4.14) and hydrophobic (eluting at ~85% acetonitrile from a reversed-phase column; Duncan 2004). It is characterized as having a "broad" spectrum of activity when challenged against other haloarchaeons (Table 5.1) but significantly, it also kills the crenarchaeal hyperthermophile *S. solfataricus* (Haseltine et al. 2001). *Sulfolobus solfataricus* mutants resistant to halocin A4 have been isolated (Haseltine et al. 2001), suggesting that there may be a common archaeal-specific target site shared by *Sulfolobus* and haloarchaeal cells sensitive to this halocin.

5.2.5.5 Halocin C8

Groundbreaking discoveries in the halocin field have been made by studying various aspects of halocin C8 produced by *Halobacterium* strain AS7092, isolated from the Great Chaidan Salt Lake, China (Li et al. 2003; Sun et al. 2005). It is the largest member of the microhalocin family (7.44 kDa, 76 amino acids) and is cysteine-rich, containing 10 cysteine residues. Halocin C8 is processed from the C-terminal end of a 283 amino acid preproprotein (called ProC8). The amino terminus contains a Tat leader sequence followed by a 207 amino acid, hydrophilic protein that confers immunity (called HalI). These are the first examples of both halocin immunity, and of an immunity protein and an antimicrobial peptide encoded in a single gene. In vitro, both unprocessed ProC8 and HalI containing the Tat leader sequence conferred immunity. HalI is associated with the membrane fraction of *Halobacterium* strain AS7092 (anchored by the Tat sequence?), and is thought to function by sequestering HalC8. In addition, heterologous expression of the gene sequence encoding HalI (named *halI* and under control of the bacterio-opsin promoter) in the HalC8-sensitive strain *Har. hispanica* also conferred immunity.

5.2.6 Protein Halocins (> 10 kDa)

5.2.6.1 Halocin H1

Halocin H1 (HalH1) is produced by *Hfx. mediterranei* M2a (formerly strain Xia3), originally isolated from a solar saltern near Alicante, Spain

(Rodriguez-Valera et al. 1982). It is a 31 kDa protein that is heat-labile (Platas 1995; Platas et al. 1996) and requires a minimum of 5% (w/v) NaCl to retain activity (Platas et al. 2002). Platas et al. (1996) determined that the nutrient source contained within the growth medium was the most important parameter influencing halocin production; growth in N-Z amine E yielded 1,280 AU/ml of halocin activity, while all other nutrients tested resulted in lower halocin production, ranging from 0–320 AU/ml of activity. The specific mode of action of HalH1 is unknown, but it appears to affect membrane permeability of sensitive cells (Platas 1995).

5.2.6.2 Halocin H4

Halocin H4 (HalH4) is produced by *Hfx. mediterranei* R4 (ATCC 33500), also originally isolated from a solar saltern near Alicante, Spain (Rodriguez-Valera et al. 1982). It was the first halocin discovered (Rodriguez-Valera et al. 1982), and has been fully characterized at both the protein and genetic levels (Meseguer and Rodriguez-Valera 1985, 1986; Cheung et al. 1997; Perez 2000). The molecular mass of HalH4 initially was determined to be approximately 28 kDa, using gel filtration and SDS-PAGE (Meseguer and Rodriguez-Valera 1985). However, once the *halH4* gene was cloned and the amino acid sequence determined by Edman degradation, the molecular mass of the mature HalH4 protein was calculated to be 34.9 kDa (359 amino acids), processed from a preprotein of 39.6 kDa (Cheung et al. 1997). The preprotein contains a 46 amino acid N-terminal Tat signal sequence (atypically long; Eichler 2000) important in translocation of the protein across the membrane (Cheung et al. 1997). How, when, and where the signal sequence is removed from the preprotein is unknown. The mature halocin also contains a 32 amino acid hydrophobic region in the middle of the protein sequence, which may be functionally important (e.g., in binding to the target site; Shand et al. 1999). The *halH4* gene consists of a 1,077-bp open reading frame encoding the 359 amino acid preprotein (Cheung et al. 1997). Cheung et al. (1997) concluded that expression of the *halH4* gene, in addition to being regulated at the level of transcription, must also be regulated post-transcriptionally. Halocin H4 is an archaeolytic halocin, described by Meseguer and Rodriguez-Valera (1986) as having "single-hit kinetics" similar to halocin H6/H7 (see Sect. 5.2.5.3). Halocin H4 adsorbs to sensitive *Hbt. salinarum* cells where it appears to disrupt membrane permeability, resulting in an ionic imbalance and leading to cell lysis. Examination of halocin activity showed sensitive cells became swollen and spherical in the presence of HalH4 (Meseguer and Rodriguez-Valera 1986), indicating that its primary target is localized in the membrane (Rodriguez-Valera et al. 1982; Meseguer and Rodriguez-Valera 1986). However, experiments to elucidate the specific target site have not revealed the actual target (Meseguer et al. 1995).

5.3 Biotechnology of Halocins

The potential of halocins as chemotherapeutic agents active against human or animal pathogens has been unrealized, but is potentially vast, given the hundreds of different halocins reported to exist versus the number of halocins actually characterized. Halocin H7, however, has been shown to inhibit the Na^+/H^+ antiporter (aka "exchanger") in both haloarchaea (Meseguer et al. 1995) and in a dog model (Alberola et al. 1998). The latter is significant, in that this halocin may serve as treatment to reduce injury caused when ischemic transplanted organs are reperfused (e.g., by reducing infarct size and the number of ectopic beats in a heart transplant; Alberola et al. 1998). The basis of this biomedical application was the discovery of the mechanism of action of halocin H7. Consequently, applications for other halocins will also hinge on the discovery of their mechanisms of action.

The *halI* gene may serve as a useful selectable marker especially for haloarchaea that require the highest levels of NaCl for optimal growth (Sun et al. 2005). Similarly, if the *S. solfataricus* gene that carries a mutation for resistance to halocin A4 can be isolated, it too might serve as a selectable marker for these crenarchaeal hyperthermophiles (Haseltine et al. 2001; O'Connor and Shand 2002).

5.4 Sulfolobicins

The archaeocins produced by *Sulfolobus* are entirely different from halocins, since their activity is predominantly associated with the cells and not the supernatant (Prangishvili et al. 2000). Prangishvili et al. (2000) were the first to isolate and characterize these proteinaceous toxins, which they called "sulfolobicins", in keeping with bacteriocin nomenclature. Provisionally, the producer strain has been named "*Sulfolobus islandicus*". Screening for sulfolobicin activity involves spotting samples of exponentially growing "*S. islandicus*" cells onto lawns of the sensitive strain *S. solfataricus* P1. Following incubation, nearly clear zones with sharp borders are generated, the size of the zone of inhibition being inversely proportional to the concentration of sensitive cells in the lawn. To date, the spectrum of sulfolobicin activity appears to be restricted to other members of the sulfolobales: the sulfolobicin inhibited *S. solfataricus* P1, *S. shibatae* B12, and six non-producing strains of "*S. islandicus*". Activity appears to be archaeocidal but not archaeolytic. It does not inhibit *S. acidocaldarius* DSM639, nor does purified sulfolobicin from strain HEN2/2 inhibit *Hbt. salinarum* R1 or *Escherichia coli* (Prangishvili et al. 2000).

Unlike halocins, sulfolobicins are not secreted into the culture medium in any significant quantity, and classical inducing agents (UV light, temperature and pH shifts, and exposure to sensitive cells) used to increase secretion have not been successful (Prangishvili et al. 2000). Analysis of sulfolobicin activity in a 500 ml culture revealed that 30 times more activity can be purified from the cell pellet than from the culture supernatant. To visualize activity in culture supernatants, the supernatant from stationary phase cultures had to be concentrated 100-fold, either by precipitation or centrifugation, before any activity was detected when spotted onto a lawn of sensitive cells.

Extracellular activity is associated with spherical particles 90 to 180 nm in diameter. These particles are present in a ratio of 1:100 cells, and are also produced by strains that do not make sulfolobicin. When purified using CsCl density gradient centrifugation, these particles form a discrete band with a density of approximately 1.29 g/ml. Electron micrographs of this material revealed an inner core with a surrounding layer having a periodicity of 22 nm, the same as the lattice constant of the *Sulfolobus* S-layer (Prangishvili et al. 2000).

Purification of sulfolobicin involves harvesting cells from late stationary phase, sonicating them, collecting the resultant cell ghosts by high-speed centrifugation, and releasing the sulfolobicin with Triton X-100. Activity elutes in the range of 30 to 40 kDa on size exclusion chromatography, in contrast to 20 kDa on SDS-PAGE. These data suggest that this archaeocin may aggregate (Prangishvili et al. 2000). Activity of purified sulfolobicin remains stable after 6 months at 4°C or 5 days at 85°C. Enzymatic treatment with α-amylase, α- and β-glucosidases, phospholipase C, and lipoprotein lipase had no effect on activity. However, treatment with pronase E, proteinase K, and trypsin completely destroyed activity, indicating activity is associated with a proteinaceous component (Prangishvili et al. 2000).

Sulfolobicins exhibit some classical bacteriocin characteristics, as they are proteinaceous and are directed against strains that are closely related to the producer. Although some of the producer strains contain conjugative plasmids, neither sulfolobicin production nor immunity can be transferred to non-producer strains, suggesting that the genes for these traits may be located on the chromosome. Although evidence suggests that sulfolobicins remain bound to cells or associated with S-layer-coated vesicles, it does not exclude the possibility that an undetectable amount of sulfolobicin may leak out from cells or vesicles into the surrounding medium. Indeed, such a scenario could account for the generation of large zones of inhibition on solid medium where the concentration of free sulfolobicin would remain more localized and high. This phenomenon also is seen with cell-bound bacteriocins (Prangishvili et al. 2000).

References

Alberola A, Meseguer I, Torreblanca M, Moya A, Sancho S, Polo B, Soria B, Such L (1998) Halocin H7 decreases infarct size and ectopic beats after mycardial reperfusion in dogs. J Physiol 509P, 148P

Baliga NS, Bonneau R, Facciotti MT, Pan M, Glusman G, Deutsch EW, Shannon P, Chiu Y, Weng RS, Gan RR, Hung P, Date SV, Marcotte E, Hood L, Ng WV (2004) Genome sequence of *Haloarcula marismortui*: a halophilic archaeon from the Dead Sea. Genome Res 14:2221-2234. Erratum in: Genome Res 14:2510

Cheung J, Danna KJ, O'Connor EM, Price LB, Shand RF (1997) Isolation, sequence, and expression of the gene encoding halocin H4, a bacteriocin from the halophilic archaeon *Haloferax mediterranei* R4. J Bacteriol 179:548-551

De Castro RE, Maupin-Furlow JA, Giménez MI, Seitz MKH, Sánchez JJ (2006) Haloarchaeal proteases and proteolytic systems. FEMS Microbiol Rev 30:17-35

Duncan JL (2004) Haloarchaeal growth physiology, characterization of halocin A4 and cloning *tbp* and *tfb* genes. Masters Thesis, Northern Arizona University, Flagstaff, AZ

Ebert K, Goebel W, Rdest U, Surek B (1986) Genes and genome structures in the archaebacteria. Syst Appl Microbiol 7:30-35

Eichler J (2000) Archaeal protein translocation: crossing membranes in the third domain of life. Eur J Biochem 267:3402-3412

Falb M, Pfeiffer F, Palm P, Rodewald K, Hickmann V, Tittor J, Oesterhelt D (2005) Living with two extremes: conclusions from the genome sequence of *Natronomonas pharaonis*. Genome Res 15:1336-1343

Gratia A (1925) Sur un remarquable exemple d'antagonisme entre deux souches de Colibacille. C R Soc Biol 93:1040-1041

Haseltine C, Hill T, Montalvo-Rodriguez R, Kemper SK, Shand RF, Blum P (2001) Secreted euryarchaeal microhalocins kill hyperthermophilic crenarchaea. J Bacteriol 183:287-291

Kirkup BC, Riley MA (2004) Antibiotic-mediated antagonism leads to a bacterial game of rock-paper-scissors *in vivo*. Nature 428:412-414

Kis-Papo T, Oren A (2000) Halocins: are they involved in the competition between halobacteria in saltern ponds? Extremophiles 4:35-41

Lenski RE, Riley MA (2002) Chemical warfare from an ecological perspective. Proc Natl Acad Sci USA 99:556-558

Li Y, Xiang H, Liu J, Zhou M, Tan H (2003) Purification and biological characterization of halocin C8, a novel peptide antibiotic from *Halobacterium* strain AS7092. Extremophiles 7:401-407

Meseguer I, Rodriguez-Valera F (1985) Production and purification of halocin H4. FEMS Microbiol Lett 28:177-182

Meseguer I, Rodriguez-Valera F (1986) Effect of halocin H4 on cells of *Halobacterium halobium*. J Gen Microbiol 132:3061-3068

Meseguer I, Rodríguez-Valera F, Ventosa A (1986) Antagonistic interactions among halobacteria due to halocin production. FEMS Microbiol Lett 36:177-182

Meseguer I, Torreblanca M, Konishi T (1995) Specific inhibition of the halobacterial Na^+/H^+ antiporter by halocin H6. J Biol Chem 270:6450-6455

Ng WV, Kennedy SP, Mahairas GG, Berquist B, Pan M, Shukla HD, Lasky SR, Baliga NS, Thorsson V, Sbrogna J, Swartzell S, Weir D, Hall J, Dahl TA, Welti R, Goo YA, Leithauser B, Keller K, Cruz R, Danson MJ, Hough DW, Maddocks DG, Jablonski PE, Krebs MP, Angevine CM, Dale H, Isenbarger TA, Peck RF, Pohlschroder M, Spudich JL, Jung K-H, Alam M, Freitas T, Hou S, Daniels CJ, Dennis PP, Omer AD, Ebhardt H, Lowe TM, Liang P, Riley M, Hood L, DasSarma S (2000) Genome sequence of *Halobacterium* species NRC-1. Proc Natl Acad Sci USA 97:12176-12181

O'Connor EM (2002) Purification and characterization of microhalocin R1 from *Halobacterium salinarum* GN101. Doctoral Dissertation, Northern Arizona University, Flagstaff, AZ

O'Connor EM, Shand RF (2002) Halocins and sulfolobicins: The emerging story of archaeal protein and peptide antibiotics. J Indust Microbiol Biotechnol 28:23-31

Perez AM (2000) Growth physiology of *Haloferax mediterranei* R4 and purification of halocin H4. Masters Thesis, Northern Arizona University, Flagstaff, AZ

Platas G (1995) Caracterizacíon de la actividad antimicrobiana de la haloarquea *Haloferax mediterranei* Xia3. Tesis Doctoral, Universidad Autónoma de Madrid, Madrid

Platas G, Meseguer I, Amils R (1996) Optimization of the production of a bacteriocin from *Haloferax mediterranei* Xia3. Microbiología 12:75-84

Platas G, Meseguer I, Amils R (2002) Purification and biological characterization of halocin H1 from *Haloferax mediterranei* M2a. Int Microbiol 5:15-19

Prangishvili D, Holz I, Stieger E, Nickell S, Kristjansson JK, Zillig W (2000) Sulfolobicins, specific proteinaceous toxins produced by strains of the extremely thermophilic archaeal genus *Sulfolobus*. J Bacteriol 182:2985-2988

Price LB, Shand RF (2000) Halocin S8: a 36-amino-acid microhalocin from the haloarchaeal strain S8a. J Bacteriol 182:4951-4958

Rdest U, Sturm M (1987) Bacteriocins from halobacteria. In: Burgess R (ed) Protein purification: micro to macro. Alan R. Liss, New York, pp 271-278

Riley MA, Wertz JE (2002) Bacteriocin diversity: ecology and evolutionary perspectives. Biochimie 84:357-364

Robinson JL, Pyzyna B, Atrasz RG, Henderson CA, Morrill KL, Burd AM, DeSoucy E, Fogelman III RE, Naylor JB, Steele SM, Elliott DR, Leyva KJ, Shand RF (2005) Growth kinetics of extremely halophilic *Archaea* (Family *Halobacteriaceae*) as revealed by Arrhenius plots. J Bacteriol 187:923-929

Rodriguez-Valera F, Juez G, Kushner DJ (1982) Halocins: salt-dependent bacteriocins produced by extremely halophilic rods. Can J Microbiol 28:151-154

Rodríguez-Valera F, Acinas SG, Antón J (1999) Contributions of molecular techniques to the study of microbial diversity in hypersaline environments. In: Oren A (ed) Microbiology and biogeochemistry of hypersaline environments. CRC Press, Boca Raton, FL, pp 27-38

Shand RF (2006) Detection, quantification and purification of halocins: peptide antibiotics from haloarchaeal extremophiles. In: Rainey FA, Oren A (eds) Extremophiles. Elsevier/Academic Press, Amsterdam, Methods in Microbiology vol 35, pp 703-718

Shand RF, Price LB, O'Connor EM (1999) Halocins: protein antibiotics from hypersaline environments. In: Oren A (ed) Microbiology and biogeochemistry of hypersaline environments. CRC Press, Boca Raton, FL, pp 295-306

Soppa J, Oesterhelt D (1989) *Halobacterium* sp. GRB: a species to work with? Can J Microbiol 35:205-209

Sun C, Li Y, Mei S, Lu Q, Zhou L, Xiang H (2005) A single gene directs both production and immunity of halocin C8 in a haloarchaeal strain AS7092. Mol Microbiol 57:537-549

Tagg JR (1992) Bacteriocins of Gram-positive bacteria: An opinion regarding their nature, nomenclature and numbers. In: James R, Lazdunski C, Pattus F (eds) Bacteriocins, microcins and lantibiotics. Springer, Berlin Heidelberg New York, NATO ASI Ser H, Cell Biol vol 65, pp 33-35

Torreblanca M, Rodriguez-Valera F, Juez J, Ventosa A, Kamekura M, Kates M (1986) Classification of non-alkaliphilic halobacteria based on numerical taxonomy and polar lipid composition, and description of *Haloarcula* gen. nov. and *Haloferax* gen. nov. Syst Appl Microbiol 8:89-99

Torreblanca M, Meseguer I, Rodríguez-Valera F (1989) Halocin H6, a bacteriocin from *Haloferax gibbonsii*. J Gen Microbiol 135:2655-2661

Torreblanca M, Meseguer I, Ventosa A (1994) Production of halocin is a practically universal feature of archaeal halophilic rods. Lett Appl Microbiol 19:201-205

Woese CR, Kandler O, Wheelis ML (1990) Towards a natural system of organisms: proposal for the domains Archaea, Bacteria and Eucarya. Proc Natl Acad Sci USA 87:4576-4579

6 The Ecological and Evolutionary Dynamics of Model Bacteriocin Communities

BENJAMIN KERR

No one knows whether death, which people fear to be the greatest evil, might not be the greatest good.

Plato, *The Apology of Socrates*

Summary

The use of model laboratory communities, model organisms, and mathematical models has deeply enriched our understanding of the causes and consequences of toxin production in bacteria. In particular, such models have provided much insight into the dynamics of microbial communities with toxin producers. Both experimental and theoretical approaches have suggested that population structure can be critical to the initial invasion of a toxin-producing strain. Furthermore, spatial structure may play a central role in the maintenance of diverse assemblages of toxic and non-toxic strains. Models have also revealed some counter-intuitive predictions, such as the evolution of competitive restraint in communities with toxin-sensitive, toxin-resistant, and toxin-producing bacteria. Toxin production itself is a dramatic form of niche construction, where producing strains alter the chemical nature of their surroundings. Such modification feeds back to affect the ecology and evolution of all community members. Models have helped greatly to clarify the effects of this feedback.

6.1 Introduction

Allelopathy, defined as the suppression or death of one organism due to the toxic chemicals excreted by another organism, is a ubiquitous phenomenon within microbial communities. In bacterial assemblages, the agents of allelopathic interaction are the bacteriocins. Bacteriocins are narrow-spectrum antimicrobial proteins found within nearly every major lineage of Bacteria

---

Department of Biology, University of Washington, Box 351800, Seattle, WA 98195, USA, e-mail: kerrb@u.washington.edu

(Riley and Wertz 2002a, 2002b). Given that bacteriocinogenic (toxin-producing) strains kill closely related non-producing strains, bacteriocins are commonly interpreted to be anticompetitor compounds (Riley 1998; Riley and Gordon 1999). Over the past few decades, there has been much interest in exploring the microbial dynamics of toxic consortia (Adams et al. 1979; Chao and Levin 1981; Levin 1988; Frank 1994; Tan and Riley 1996; Durrett and Levin 1997; Iwasa et al. 1998; Gordon and Riley 1999; Pagie and Hogeweg 1999; Nakamaru and Iwasa 2000; Czárán et al. 2002; Kerr et al. 2002; Czárán and Hoekstra 2003; Kirkup and Riley 2004). Some of these studies have shown that Socrates' insight carries particular salience for communities with bacteriocinogenic members – allelopathy may play a critical role in maintaining diversity in these systems (Durrett and Levin 1997; Pagie and Hogeweg 1999; Czárán et al. 2002; Lenski and Riley 2002; Kerr et al. 2002).

The best-studied case of microbial allelopathy is found in the bacterium *Escherichia coli*, which possesses many toxic strains. In *E. coli*, the gene encoding the toxin (termed a colicin) is housed on a plasmid along with a constitutively expressed immunity gene (conferring protection against the action of the colicin) and a lysis gene (usually expressed under conditions of stress, causing lysis of the cell and subsequent release of the colicin; James et al. 1996). Thus, in *E. coli* (as well as other Gram-negative bacterial species) bacteriocinogenic cells die in the process of releasing their toxins. A plausible interpretation is that the lethal release of toxins kills non-producing competitors, promoting the spread of remaining clone mates that carry the plasmid encoding immunity to the toxin. However, under precisely what circumstances would such lethal production evolve? And in communities with producers, what are the expected population-level consequences?

Models have proven extremely useful in answering such questions. Indeed, much of the current understanding of bacteriocin systems has come through the use of models, taken broadly to include model organisms (such as *E. coli*), model laboratory communities, and theoretical models. The foundational studies of bacteriocin-mediated community dynamics were done with *E. coli* in experimental microcosms (Adams et al. 1979; Chao and Levin 1981), and laboratory communities have continued to provide insight, both in vitro (Tan and Riley 1996; Riley and Gordon 1996; Gordon and Riley 1999; Wiener 2000; Kerr et al. 2002; Massey et al. 2004) and in vivo (Kirkup and Riley 2004; Massey et al. 2004). Such studies are often motivated by one of two related questions: how does toxin production arise? And how does toxicity influence the dynamics of the community?

The latter question has been targeted by several theoretical biologists studying bacteriocin systems. Given the large number of different bacteriocinogenic constituents in natural microbial communities, theoreticians have been nearly singularly motivated by providing mechanisms of diversity maintenance. In the process, theoretical biologists have brought a varied analytical and computational set of tools to the task, including systems of ordinary differential equations (Durrett and Levin 1997; Gordon and Riley 1999), reaction-diffusion equations (Frank 1994; Nakamaru and Iwasa 2000), pair approximation (Iwasa et al. 1998), configuration field approximation (Czárán and Hoekstra 2003), and agent-based simulation (Durrett and Levin 1997; Pagie and Hogeweg 1999; Kerr et al. 2002; Czárán and Hoekstra 2003).

In this chapter, I will review the contributions of models to a deeper understanding of the causes and consequences of microbial allelopathy. The study of bacteriocin communities has benefited tremendously from a dialogue between theorists and empiricists. I will discuss some of the ways in which the theory has been inspired by and has, in turn, inspired experimental work. Finally, I will identify a few areas where the continued interaction between theoretical work, experimental work and natural history may produce deeper understanding. The following sections are organized according to structural complexity of the model bacteriocin community – starting with the simplest single-producer communities and ending with multiple-producer communities.

6.2 Dynamics in Two-Strain Communities: Getting over the Hump

The simplest bacteriocin community consists of two players: a strain producing the toxin and a strain sensitive to the toxin. For bacterial species such as *E. coli*, toxin production can be costly due to constitutively expressed immunity, plasmid carriage and lethality of production (Riley and Gordon 1999; Riley and Wertz 2002a, 2002b). This cost has been demonstrated in the laboratory, where the producer has a lower growth rate or a higher mortality rate than the sensitive strain (Adams et al. 1979; Chao and Levin 1981; Tan and Riley 1996, but see Dykes and Hastings 1997 for a discussion of Gram-positive producers). Given this cost, if the sensitive strain and producing strain were growing in two separate flasks, then the sensitive strain has the edge. But what happens when both strains are mixed in the same flask?

In well-mixed conditions (such as a shaken flask or a chemostat), bacteriocins released by a producer are evenly distributed throughout the entire community. This means that the per capita effect of the toxin on the pool of sensitive cells scales with the number of toxin producers – the more producers, the higher the per capita mortality rate for the sensitive strain. If there are very few producers in the community, then the impact of the bacteriocin on the sensitive pool will be minimal. In such a case, there will be a net growth advantage for the sensitive strain (as production is costly), and the sensitive strain can displace the producer. Alternatively, if toxin-producing cells are common, then the impact on the sensitive pool can be pronounced. Despite the intrinsic cost of toxin production, a high density of the producer can create a heavy *extrinsic* cost in net growth rate for the sensitive strain. If the bacteriocin is sufficiently toxic, the producing strain can displace the sensitive strain. Under such cases, there is "strength in numbers" under mass-action conditions: above some threshold, the producer can administer enough poison to overburden its sensitive competitor.

A mathematical treatment of the competition between the producer and sensitive strain is given in Box 1. If the producer is sufficiently toxic, the community is bistable: either the producer excludes the sensitive strain or vice versa, depending on initial conditions (Levin 1988; Frank 1994; Durrett and Levin 1997; Iwasa et al. 1998). This bistability has been confirmed in the laboratory: under well-mixed conditions and constant initial density, a producer displaces its sensitive competitor *only* if above a critical frequency (Adams et al. 1979; Chao and Levin 1981). So, invading producers do have a proverbial hump to get over.

Box 1: A mass-action model of a producer strain and a sensitive strain

Durrett and Levin (1997) use the following system of differential equations to model the community dynamics of a sensitive strain (with density s) and producer strain (with density p):

$$\frac{ds}{dt} = \beta_s (1 - s - p) s - (\delta_s + \gamma p) s \qquad \text{(B1.1)}$$

$$\frac{dp}{dt} = \beta_p (1 - s - p) p - \delta_p p \qquad \text{(B1.2)}$$

where β_s and β_p are the birth rates, and δ_s and δ_p are the death rates of the sensitive and producer strains, respectively, and γ measures the per capita toxic effect of producers on the sensitive strain. We assume that

$$\beta_s > \delta_s \qquad \text{(B1.3)}$$

$$\beta_p > \delta_p \qquad \text{(B1.4)}$$

That is, each strain's reproductive gains outstrip its losses to intrinsic death. When alone, the carrying capacity of strain i is $1 - \delta_i/\beta_i$ (the carrying capacity approaches the maximum of unity as $\delta_i \to 0$ or $\beta_i \to \infty$). We also require that

$$1 - \frac{\delta_s}{\beta_s} > 1 - \frac{\delta_p}{\beta_p} \qquad \text{(B1.5)}$$

That is, the sensitive strain has a higher carrying capacity than the producer when each is in isolation.

The current state of the two-strain community can be expressed as a point (p, s) on the two-dimensional p–s plane (see Fig. 6.B1a, d). Tracking community behavior amounts to following the trajectory of this point over time. The point moves according to Eqs. (B1.1) and (B1.2). One way to get some insight into the point's movement is to draw zero net growth

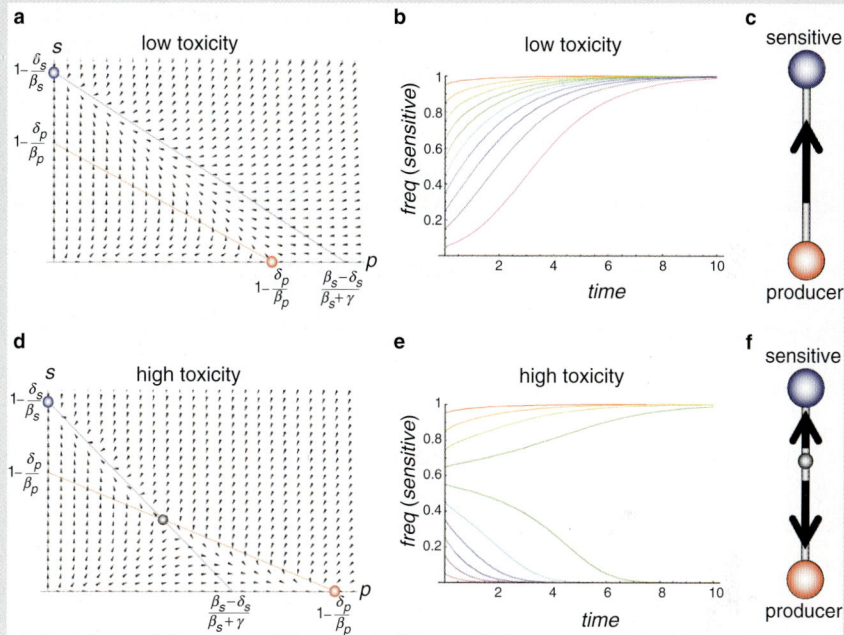

Fig. 6.B1 Exclusion or bistability in a two-strain community. **a** The isoclines for the sensitive (*blue*) and producer (*red*) strains are shown in the p–s plane. The *arrows* give the flow of a point describing the densities of the two strains. When the producer is insufficiently toxic ($\gamma < \gamma_c$), the isoclines do not cross. From nearly all starting positions, the "community point" moves to the equilibrium on the s axis (given by the *blue sphere*). That is, the sensitive strain displaces the producer. **b** Here, we see the same dynamics expressed as the frequency of the sensitive strain over time. Despite the starting conditions, the sensitive type fixes (that is, it approaches a frequency of 1). **c** A symbolic representation of the community dynamics (see Fig. 6.2). The *arrow* pointing from the producer node to the sensitive node indicates that the sensitive strain will outcompete the producer under any starting conditions. **d** When the producer is sufficiently toxic ($\gamma > \gamma_c$), the isoclines cross and a new internal equilibrium (the *gray sphere*) is introduced. This new equilibrium is unstable. In this community, the initial strain densities become important – if the producer is sufficiently abundant relative to the sensitive strain, then the producer will displace the sensitive strain and vice versa. This is a bistable system where both edge equilibria (the *red* and *blue spheres*) are locally stable. **e** Now, the sensitive strain fixes only if frequent enough – otherwise, it goes extinct (and the producer fixes). **f** The symbolic representation shows *arrows* pointing to each node with an unstable internal node in between

Box 1: Continued

isoclines for each strain (the isocline for each strain is a curve where it does not change its density).

The isocline for the sensitive strain (found by setting $\frac{ds}{dt} = 0$) is a line in the p–s plane:

$$s = \left(1 - \frac{\delta_s}{\beta_s}\right) - \left(1 + \frac{\gamma}{\beta_s}\right)p \quad (B1.6)$$

Similarly, the isocline for the producer strain (found by setting $\frac{dp}{dt} = 0$) is:

$$s = \left(1 - \frac{\delta_p}{\beta_p}\right) - p \quad (B1.7)$$

If the point in the plane (giving the strains' densities) is above the sensitive's isocline, then it must move downward (because the density of the sensitive strain is on the vertical axis, and if $s > 0$, then $s > (1 - \delta_s/\beta_s) - (1 + \gamma/\beta_s)p \Rightarrow ds/dt < 0$). On the other hand, if the point is below the sensitive's isocline, it must move upward. Simultaneously, if a point is above the producer's isocline, then it must move leftward (because the density of the producer is on the horizontal axis, and if $p > 0$, then $s > (1 - \delta_p/\beta_p) - p \Rightarrow dp/dt < 0$). By contrast, if the point is below the producer's isocline, then it must move rightward.

Therefore, the positioning of the isoclines (whether and how they cross) can yield important information about community dynamics. In this two-strain system, there is a critical toxicity of the producer

$$\gamma_c = \frac{\beta_s \delta_p - \beta_p \delta_s}{\beta_p - \delta_p} \quad (B1.8)$$

By assumptions (B1.3), (B1.4) and (B1.5), $\gamma_c > 0$. If $\gamma < \gamma_c$ (that is, the producer is not very toxic), then the isoclines do not cross in the positive quadrant of the p–s plane (see Fig. 6.B1a, where the sensitive isocline is in blue and the producer isocline is in red). The arrows in Fig. 6.B1a trace out the potential movement of a point giving the strain densities. Note that the arrows cut the blue line horizontally (because vertical movement of the point corresponds to changes in the sensitive strain, and the sensitive strain does not change its density on its isocline), and the arrows cut the red line vertically (because horizontal movement of the point corresponds to changes in the producer, and the producer does not change its density on its isocline).

As the figure shows, from nearly any starting condition, the community moves to the boundary equilibrium $(0, 1 - \delta_s/\beta_s)$ given by the blue sphere, where the sensitive strain excludes the producer. There also exists an unstable equilibrium $(1 - \delta_p/\beta_p, 0)$ given by the red sphere (introducing sensitive cells into a population of producers at the producer carrying capacity would lead to the exclusion of producers by the invading sensitive strain). In Fig. 6.B1b, we see that the frequency of the sensitive strain approaches unity despite starting conditions (Fig. 6.B1c shows this behavior schematically). Thus, without sufficient toxicity, the producer always goes extinct in head-to-head competition.

If the toxicity of the producer is above the critical level ($\gamma > \gamma_c$), then both boundary equilibria become locally stable and the isoclines cross at the point $((\beta_s \delta_p)/(\beta_p \gamma) - \delta_s/\gamma, 1 + \delta_s/\gamma - (1 + \beta_s/\gamma) \delta_p/\beta_p)$ in the positive quadrat. This point is an unstable equilibrium (this can be shown locally using linear stability analysis; see the Appendix). From most starting positions, either the sensitive strain displaces the producer or vice versa (see Fig. 6.B1d). Thus, initial community composition becomes important in determining which strain dominates. Generally, if sufficiently abundant, the producer displaces the sensitive strain, otherwise it goes extinct. This is shown in Fig. 6.B1e (and schematically in Fig. 6.B1f). This bistability was demonstrated in vitro with *E. coli* (Adams et al. 1979; Chao and Levin 1981).

Both the mathematical and empirical results discussed above depend critically on the assumption of a well-mixed community. In an ingenious experiment, Chao and Levin (1981) competed a producer and a sensitive strain of *E. coli* in two different habitats: (1) a well-mixed broth-filled flask and (2) an agar-filled Petri dish. They found bistability in the stirred flask (the producer displaced the sensitive strain only when above a critical threshold). However, the producer *always* displaced the sensitive strain in the spatially structured dish (i.e., even if the producer was extremely rare, it displaced the sensitive strain). So, spatial structure had effectively leveled the producer's hump. Why might this be?

Consider a scenario in which producers are very rare in the Petri dish. In such a spatially structured environment, the toxin released by a producer is not distributed to all members of the community. Rather, the sensitive neighbors of producers experience a disproportionately high dose of the toxin. As a consequence, the mortality rate of sensitive cells near toxin-producing cells is higher than that of the average sensitive cell. Given that reproduction also occurs locally, the space liberated near a producer (through the elevated deaths of sensitive cells) is disproportionately available to toxin-producing cells. In this way, small clumps of producers can "toxically clear-cut" sensitive cells at their periphery and radiate outward into a sea of sensitivity (see Fig. 6.1). Since Chao and Levin's pioneering study, the loss of bistability in structured bacteriocin communities has been demonstrated theoretically through lattice-based simulation (Durrett and Levin 1997) and pair approximation (Iwasa et al. 1998).

Given the above theoretical and experimental results, the prospect for the long-term coexistence of producer and sensitive strains looks fairly grim. In well-mixed habitats, one of the two strains is predicted to displace the other,

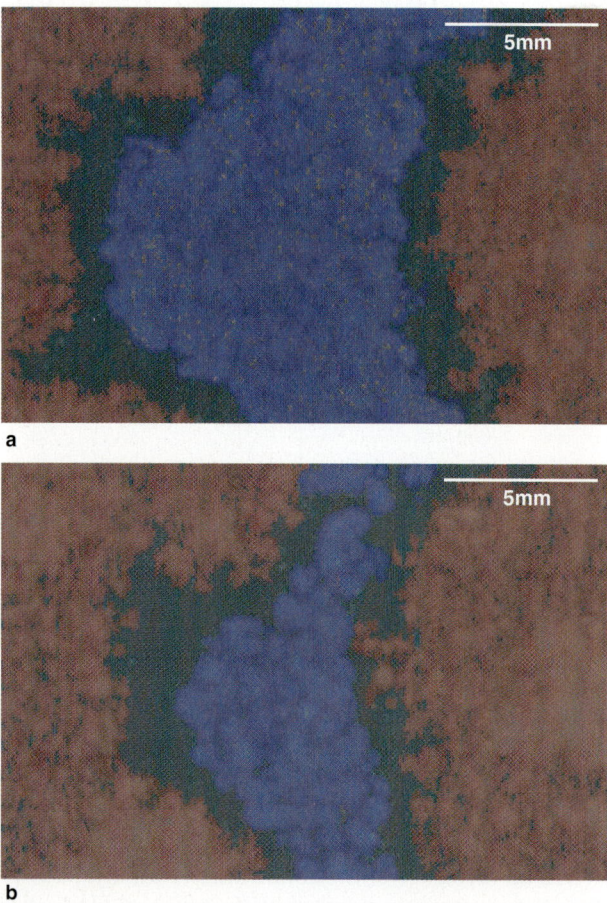

Fig. 6.1 The spatial advance of a producer strain. **a** A photograph of a bacterial community growing on the surface of an agar plate. The bacterial patches highlighted in *red* are producers (E2 colicinogenic *E. coli*) and the bacterial patches highlighted in *blue* are sensitive to the bacteriocin. The cell-free areas between the strains are zones of inhibition, where diffused toxin from the producer has prevented growth of the sensitive strain. **b** A photograph of the same field taken 24 h later (a velvet transfer of the community onto a new agar plate allowed for further growth without disrupting the spatial configuration). The producer patches have closed in on both sides of the sensitive patch. In this way, the producer (which grows to lower density on agar plates) can displace the sensitive strain through local toxic killing

depending on initial conditions. The addition of spatial structure simply tips the scales in favor of the producer. Are there circumstances under which we would expect the two strains to coexist?

Frank (1994) took a reaction-diffusion approach to modeling this two-strain system. He showed that if there is spatial heterogeneity in resource concentration, then both strains can stably coexist. In his model, toxin-producing cells inhabit resource-rich areas (where competition for resources is muted), while sensitive cells dominate the resource-poor areas (where resource competition is intense). Ultimately, Frank's model explains diversity by invoking underlying environmental heterogeneity. Although such spatial heterogeneity is not only plausible but probable, Frank's model does stimulate the following question: is it possible to maintain diversity even in a spatially homogeneous system?

In a recent paper, Czárán and Hoekstra (2003) demonstrate that the answer to this question is "yes". Their model assumes that the microbial community is distributed across many sites; collectively, the sites comprise a "metapopulation". Each site has the same properties (i.e., there is no underlying heterogeneity in this model), and the microbes are assumed to disperse among sites. If a site is simultaneously colonized by both sensitive cells and toxin-producing cells, then the producers will exclude the sensitive cells. However, the authors assume that the fast-growing sensitive cells hit high density (while the producer population is still at low numbers) before going locally extinct (as the producer increases to high density). While at high density, the sensitive cells migrate to other sites (whereas the low density producers do not). Thus, even though the fate of any sensitive strain is local extinction at a site through the toxic killing of a colonizing producer, the sensitive strain can nonetheless persist by embracing a nomadic lifestyle. As long as empty sites are continually being generated (i.e., there is some probability that a community at any given site will crash), the rapidly colonizing sensitive strain can persist globally. Such a model might be especially relevant for explaining diversity in bacteriocinogenic enterics (such as species of *Citrobacter*, *Enterobacter*, *Escherichia*, *Hafnia*, *Klebsiella*, *Serratia*, etc.), where the intestinal tracts of multiple hosts form a metapopulation.

Another explanation of the coexistence of producer and sensitive strains relies on the presence of a third strain of bacteria. We now turn to such three-member communities.

6.3 Dynamics in Three-Strain Communities: Playing Rock–Paper–Scissors

Cells sensitive to a bacteriocin will occasionally experience mutations that render them resistant. In *E. coli*, resistance often involves loss or alteration in a membrane-associated protein that binds or translocates the toxin (James et al. 1996; Feldgarden and Riley 1998, 1999; Riley and Gordon 1999). Resistance is different than immunity. Producer immunity involves a constitutively expressed immunity protein that binds and neutralizes the producer's bacteriocin, whereas resistance is often engendered by a failure of the non-producing cell to bind or import the toxin in the first place.

Under some circumstances, the resistant strain will have a growth rate intermediate between that of sensitive and producer strains. The resistant strain may grow slower than the sensitive strain because the membrane proteins that bind or translocate bacteriocins often perform other cell functions (e.g., nutrient uptake), and thus their loss or alteration can compromise such functions. On the other hand, the resistant strain will grow faster than the producer when the costs of resistance (e.g., compromised nutrient uptake) are less than the costs of bacteriocin production (e.g., plasmid carriage, constitutive immunity, lethal synthesis). Given this ordering, a sensitive strain will outgrow a resistant strain, a resistant strain will outgrow a producer, and a sufficiently common producer can displace a sensitive type through toxic killing. Such a relationship is analogous to the children's game of rock–paper–scissors (indeed, an easy way to remember this is to look at the first letters of "*r*esistant–*p*roducer–*s*ensitive", although unfortunately, according to the first letters alone, the actual dynamic turns opposite to the *r*ock–*p*aper–*s*cissors game). This non-transitive dynamic has been found to hold for *E. coli* in vitro (Kerr et al. 2002) and in vivo (Kirkup and Riley 2004).

There has been a fair amount of theoretical interest in the dynamics of such rock–paper–scissors communities (Gilpin 1975; Durrett and Levin 1997; Riley and Gordon 1999; Nakamaru and Iwasa 2000; Kerr et al. 2002). In some non-transitive communities, the three players can coexist stably. However, under mass-action conditions, this is not the case for the resistant–producer–sensitive community. Using a system of ordinary differential equations, Durrett and Levin (1997) show that one strain always drives the other two extinct. Actually, the above resistant–producer–sensitive community is a special case of the more general Durrett and Levin model. The sensitive strain is predicted to dominate the well-mixed community (Nakamaru and Iwasa 2000). In the Appendix, we prove that sensitivity is an evolutionarily stable strategy (ESS) for a simple three-strain model.

One way to visualize the dynamics in this three-strain community is to use a de Finetti diagram. Here, a single point inside or on a triangle carries all the information to deduce the frequencies of the three strains: each of the three vertices is labeled with one of the three strains, and the frequency of each strain is given simply by the normalized distance from the point to the edge opposite to the relevant vertex. For instance, a point on the "sensitive" vertex corresponds to a community fixed for sensitive types, whereas a point on the edge connecting the "resistant" and "producer" vertices corresponds to a community without any sensitive cells, and a point in the center of the triangle corresponds to a community with equal frequencies of each strain.

In Fig. 6.2a, we give the "boundary dynamics" on a de Finetti diagram for the rock–paper–scissors game (Frean and Abraham 2001; Czárán et al. 2002). We see that any community comprised of only "rock" and "paper" fixes for paper (since paper beats rock), any community of only "paper" and "scissors" fixes for scissors, and any community of only "scissors" and "rock" fixes for rock. In Fig. 6.2b, we show the dynamics when all three players are present – and we see continued cycles. In Fig. 6.2c, we give the boundary dynamics for the resistant–producer–sensitive game (when the producer is fairly toxic). Here, we see that we do not have a simple flow from one vertex to the next on the outside of the triangle. Rather, on the "sensitive–producer" edge we have flow going in both directions – both the sensitive pool and producer pool will exclude the other when they are sufficiently frequent in a two-strain community (see Fig. 6.B1f in Box 1). This is another appearance by the bistability described above. The hump (represented as a small gray point on the sensitive–producer edge in Fig. 6.2c) has reemerged. In Fig. 6.2d, we see that the dynamics are quite different than in the strict rock–paper–scissors game – the sensitive strain excludes the others from nearly every starting condition (see Nakamaru and Iwasa 2000, and the Appendix). Do note that these models assume infinite population sizes, and often the dynamical trajectory can come very close to the producer-resistant edge of the triangle (where the sensitive pool is extremely rare). Thus, in a finite population, one might often observe extinction of the sensitive strain and consequent fixation of the resistant strain (Kerr et al. 2002).

Interestingly, when this same three-member community is spatially structured (e.g., modeled as cells occupying the points of a lattice where reproduction and interaction are localized), all three strains can coexist (Durrett and Levin 1997; Kerr et al. 2002). Due to local reproduction in a spatially structured environment, clumps of each of the three strains form, and these clumps chase one another at their boundaries. Sensitive patches chase resistant patches, resistant patches chase producer patches, and producer patches in turn chase sensitive patches. Thus, all clumps are simultaneously chasing and being chased, and the upshot of this shifting mosaic is that all strains are maintained (see Box 2). By propagating three strains of *E. coli* in a well-mixed habitat (a stirred flask) and a structured habitat (the surface of an agar plate), Kerr et al. (2002) experimentally demonstrated that spatial structure can promote the maintenance of diversity in a bacteriocin community. In a sense, spatial structure in these cases obliterates the "hump" on the sensitive–producer edge of the de Finetti diagram (Durrett and Levin 1997; Iwasa et al. 1998). In Fig. 6.2f, we see simulated dynamics from the lattice-based model described in Box 2. This behavior is much closer to the rock–paper–scissors game of Fig. 6.2b (with the caveat that the arrows flow in the opposite direction).

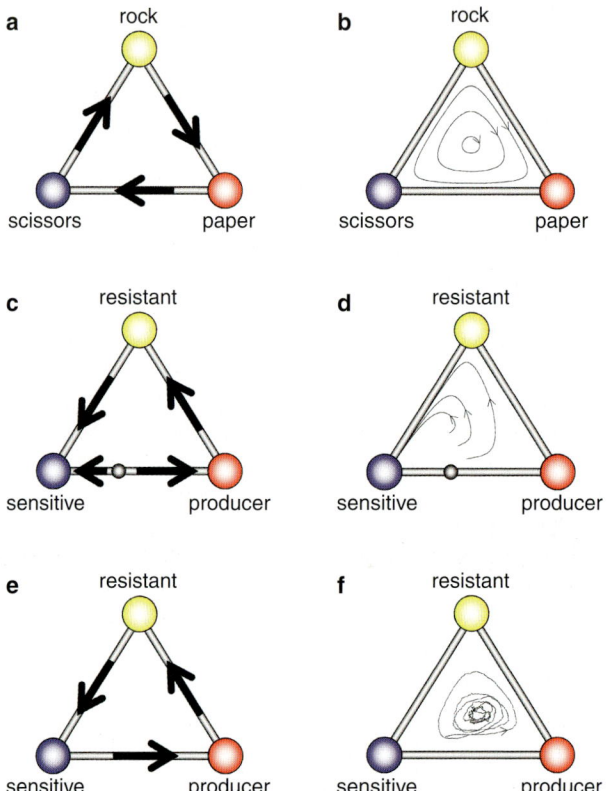

Fig. 6.2 Three-strain community dynamics. In this figure, the de Finetti diagram is used to represent changes in community composition. Each vertex of the triangle is labeled with one of the three competitors, and then the frequencies of the competitors are given by the location of a point inside or on the triangle. To find the frequency of a specific competitor, simply compute the normalized distance from the point to the edge opposite the competitor's vertex. Thus, the closer the point is to any given vertex, the more frequent is the corresponding competitor. The movement of this point traces out trajectories within the triangle that give an illustration of community dynamics. **a** This schematic gives the basic rock–paper–scissors dynamics (e.g., Frean and Abraham 2001). The *thick arrows* on the edges of the triangle give the pairwise competition outcomes. For instance, since rock beats scissors, the point giving frequencies "flows" from the scissors vertex toward the rock vertex (rock "competitors" replace scissors "competitors"). **b** When all three competitors (rock, paper, and scissors) are simultaneously present, the point is inside the triangle. The community dynamics are shown for Frean and Abraham's (2001) rock–paper–scissors model (we set their $P_r = P_s = P_p = 0.7$). The trajectories are closed loops – the frequencies of each competitor oscillate indefinitely (where the amplitude of oscillation depends on starting frequencies) and all three strains are maintained. **c** This schematic gives the resistant–producer–sensitive dynamics in a well-mixed habitat (see Appendix). The *thick arrows* indicate that the sensitive strain will outcompete the resistant strain, and the resistant strain will outcompete the producer. However, along the edge connecting the sensitive and producer vertices, there is a bistability (if producers are sufficiently common, they displace sensitive cells and vice versa – see Box 1). **d** The interior dynamics are noticeably different from the rock–paper–scissors game – the sensitive strain ends up dominating the community. **e** In the spatial version of the resistant–producer–sensitive dynamics, the rock–paper–scissors game reemerges (although the arrows flip because the pairwise competitions reverse when using the same r–p–s lettering on the triangle). **f** When all three strains are present in a finite structured lattice, the community cycles into a stable oscillating coexistence (see Box 2)

Box 2: A lattice-based three-strain model

The following approach is a slight modification of the agent-based simulations in Durrett and Levin (1997). A virtual community of sensitive cells, producers and resistant cells occupy the points of an $L \times L$ square lattice with wrap-around boundaries. To start the simulation, every point in the lattice is randomly assigned one of the following states: {S, P, R, E}, where S represents a point occupied by a sensitive cell, P is a point with a producer, R is a point with a resistant cell, and E is an empty lattice point. The community dynamics are given by an asynchronous updating scheme, in which a random sequence of focal points in the lattice are picked and the state of each focal point is changed probabilistically. For instance, an S→E transition describes the death of a sensitive cell, whereas an E→P transition describes the "birth" of a producer. The probabilities of specific state changes of a focal point depend not only on its current state, but also potentially on the states of points in its neighborhood. For instance, a sensitive cell surrounded by toxin-producing cells has a higher probability of death (i.e., the S→E transition is more likely) than an isolated sensitive cell.

By varying the size of the neighborhood, the scale of ecological processes (such as toxic interaction, competition for space, and dispersal) can be controlled. If we make the neighborhood small, then dispersal and interaction become spatially restricted. For instance, the neighborhood might be the eight nearest lattice points around a focal point (this is called a Moore neighborhood). Alternatively, we might make the neighborhood of a focal point the entire lattice (minus the focal). For such a "Global" neighborhood, the community behaves like a well-mixed system.

If we pick an empty point to update, then it becomes filled with strain i ($i \in$ {S,P,R}) with probability f_i, where f_i is the fraction of the empty point's neighborhood filled with strain i. If a point occupied by strain i is picked, then it goes to an empty state (a death event) with probability Δ_i. While Δ_P and Δ_R are assumed to be constant parameters, Δ_S is not; the death rate of a sensitive cell is assumed to increase linearly with the fraction of producers in its neighborhood:

$$\Delta_S = \Delta_{S,0} + \tau f_P \tag{B2.1}$$

where $\Delta_{S,0}$ gives the intrinsic probability of a sensitive cell's death (i.e., when there are no producers in its neighborhood) and τ measures the toxicity of the producer (τ is similar to the γ parameter in Box 1). To guarantee non-transitivity in this system, the following is assumed:

$$\Delta_{S,0} < \Delta_R < \Delta_P < \frac{\Delta_S + \tau}{1 + \tau} \tag{B2.2}$$

Box 2: Continued

In words, conditions (B2.2) simply state that there is a net growth hierarchy with the sensitive strain on the top, the resistant strain in the middle, and the producer on the bottom. However, the producer is above a critical toxic level, which yields a non-transitive competitive dynamic.

When this three-strain community is simulated using a Moore neighborhood, all three strains coexist under many different parameter settings. Because dispersal is local, clumps of the three strains form and these clumps chase one another at their boundaries – **S** clumps chase **R** clumps, **R** clumps chase **P** clumps, and **P** clumps chase **S** clumps (see Fig. 6.B2a, b). However, when a Global neighborhood is used, diversity is rapidly lost

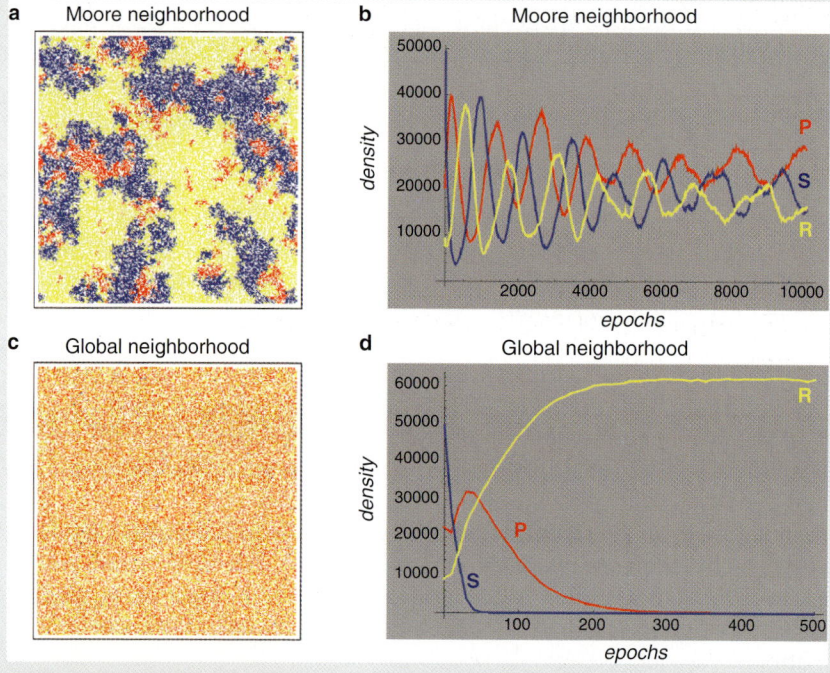

Fig. 6.B2 Lattice-based simulations. **a** A snapshot of a 300 × 300 structured lattice after 750 epochs (an epoch is $L \times L = 300 \times 300$ updates). Sensitive cells are *blue*, producers are *red* and resistant cells are *yellow*. A Moore neighborhood was used, and the parameters were $\Delta_{S,0} = 1/4$, $\Delta_P = 1/3$, $\Delta_R = 0.312$, and $\tau = 0.65$. Substantial clumping can be observed in this picture of the lattice. These clumps chase one another across the lattice according to the non-transitive dynamic. **b** The population dynamics over 10,000 epochs showing that all three strains persist for long periods of time. **c** A snapshot of a 300 × 300 unstructured lattice after 50 epochs. A Global neighborhood was used with the same parameters as for parts **a, b**. In this case, there is no spatial clumping. **d** Diversity is rapidly lost from the Global neighborhood simulation

(see Fig. 6.B2c, d). In the Global neighborhood, the toxic effects of producers are distributed globally. This can drive the sensitive strain to very low levels (unless the producer is not very toxic). Indeed, because our lattice is finite, the sensitive strain often goes extinct. Once one member of a non-transitive triplet is lost, the final competitive outcome is decided (for the same reason that a game of "rock–paper" would be much less entertaining for schoolchildren than a game of "rock–paper–scissors"). If the sensitive strain exits the community, then the resistant strain simply outcompetes the producer, and we end up with a monomorphic population.

The simulations with a Global neighborhood correspond closely to the dynamics given by the set of mean-field ordinary differential equations (see the Appendix and Fig. 6.2c, d). However, because such mathematical models assume infinite populations (and thus one can have an arbitrarily small density of sensitive cells), the sensitive strain is expected to "hang on" as the resistant displaces the producer, and eventually dominates the community. However, the outcome for the maintenance of diversity is the same: diversity is lost in the well-mixed community. Thus, population structure can be critical to coexistence. This role for spatial structure promoting diversity in a non-transitive bacteriocin community was demonstrated in vitro with *E. coli* (Kerr et al. 2002).

6.4 Evolution in Three-Strain Communities: Survival of the Weakest

Up to this point, we have considered only ecological dynamics in microbial communities. Of course, given their large population sizes and short generation times, it would be inappropriate to ignore evolution. There have been a few theoretical studies that have considered the effects of evolutionary change within a rock–paper–scissors system (Frean and Abraham 2001; Johnson and Seinen 2002). However, there has not been any detailed theoretical or experimental analysis of the evolutionary dynamics within the aforementioned resistant–producer–sensitive system.

As resistance to a bacteriocin arises readily through mutation of sensitive cells, and the cost of resistance is often variable (Feldgarden and Riley 1998, 1999), it would seem reasonable to consider the possibility that the cost of resistance can change evolutionarily. One way to model this situation is outlined in Box 3. An intuitive expectation is that the resistant strain should evolve to minimize its cost (e.g., continually lower its death rate or raise its reproductive rate). What actually occurs in simulations seems bizarre at first glance: in a spatially structured community with producer and sensitive strains, the resistant population does *not* evolve to minimize its cost! Why is this?

Box 3: An evolutionary simulation

In order to introduce evolution in the cost of resistance, we consider a small wrinkle to the lattice-based model in Box 2. Specifically, instead of fixing the probability of death of a resistant cell as a global parameter Δ_R, we allow every single resistant cell to carry its own Δ_R. Within the framework of the model, this Δ_R is the genotype of our virtual resistant cell. When a new resistant cell is "born", a mutation can occur to change the death probability. Specifically, if $\Delta_R(parent)$ is the death rate of a parent, then we assume that the death rate of an offspring is:

$$\Delta_R(offspring) = \begin{cases} \max(\min(\Delta_R(parent) + Z, \Delta_P - \varepsilon), \Delta_{S,0} + \nu) & \text{with prob. } \mu \\ \Delta_R(parent) & \text{with prob. } (1 - \mu) \end{cases}$$

(B3.1)

where μ is the probability of mutation and Z is a random variable (for instance, $Z \sim N(0, \sigma^2)$ or $Z \sim \text{Unif}(-\phi, \phi)$, where σ or ϕ relate to the amount that the death rate can change due to a single mutation). We assume that the death rate of the resistant cell must always remain intermediate between the intrinsic death rate of the sensitive strain and the death rate of the producer – the positive parameters ε and ν are taken to be small, but are nevertheless included to guarantee that, despite any evolutionary change, the non-transitive competitive structure is maintained.

When a resistant population is simulated without other competing strains, it evolves to minimize the cost of resistance (average Δ_R evolves to the minimum value in the range allowed). However, when evolution occurs in a three-strain community with local dispersal and interaction (using a Moore neighborhood), the cost of resistance does not evolve to its lowest level (see Fig. 6.3). It pays off to exercise competitive restraint in this non-hierarchical community because such restraint aids the enemy of your enemy (which, in turn, harms your enemy and thus aids you). An extremely interesting direction for future experimental work involves exploration of these counterintuitive spatial evolutionary dynamics within non-transitive systems.

The reason is given by the adage "the enemy of my enemy is my friend". In a spatially structured habitat, each strain exists as a set of clumps. These clumps are simultaneously chasing other clumps, and being chased. Now, if a mutant arises within a resistant clump that has a much reduced cost, then this mutant will start to outcompete both its fellow resistant types and any nearby producer cells. In fact, the resulting mutant clump will chase bordering producer clumps more rapidly. If the mutant has extremely low costs, then the mutant clump can chase a bordering producer clump to extinction, which puts these mutants face-to-face with a sensitive clump (an interaction in which they do not fare well). In this way, by continuing to lower costs, a resistant lineage may "improve itself to death". The fact that many such clumps simultaneously exist across a large spatial arena means that the drive within clumps to reduce the cost of resistance is checked by the enhanced probability of clump extinction. Strains that exercise restraint (i.e., maintain relatively high costs) persist by default as their less restrained cousins burn themselves out.

In Fig. 6.3, we see the maintenance of a non-minimal cost of resistance in a spatially structured three-strain community. On the other hand, if the resistant strain evolves alone in a spatially structured habitat, it does evolve to minimize its cost (Fig. 6.3). In a structured non-transitive community, a higher cost of resistance retards replacement of producers by resistant cells.

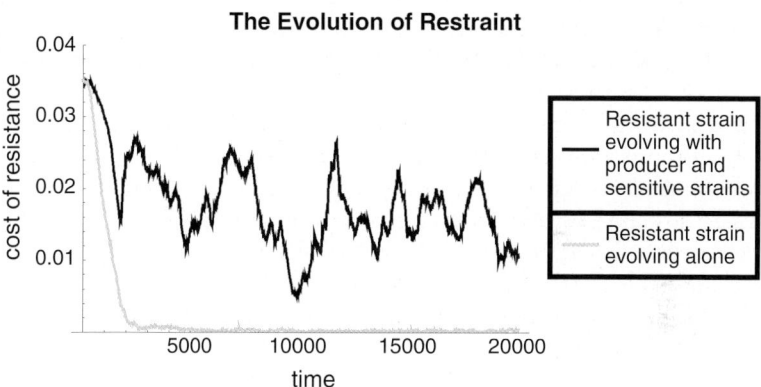

Fig. 6.3 The evolution of competitive restraint. Shown are the results of a lattice-based simulation (described in Box 2) allowing for the resistant strain to evolutionarily change the cost of resistance (see Box 3). The parameters used are $\Delta_{S,0} = 1/4$, $\Delta_P = 1/3$, $\tau = 0.55$, $\varepsilon = 0.004$, $\nu = 0.025$, $\mu = 0.001$, and $Z \sim \text{Unif}(-0.02, 0.02)$. The death rate of the resistant strain (Δ_R) can evolve. The average cost of resistance is simply $CR = \bar{\Delta}_R - \Delta_{S,0}$ where $\bar{\Delta}_R$ is the average death rate of the evolving resistant strain. The minimum value that CR can obtain is ν. A proxy for cost of resistance is $CR' = CR - \nu$. All else being equal, the resistant strain is expected to evolve to minimize its cost (i.e., we expect $CR' \to 0$). The *black trajectory* is the average cost of resistance (CR') in a 300×300 square lattice with a Moore neighborhood, where the resistant strain shares the lattice with the producer and the sensitive strain. Here, we see that the cost of resistance does not evolve to its minimum, but rather remains at higher levels (that is, the average death rate of the resistant strain is evolutionarily maintained at a value higher than its obtainable minimum). As a control, the *gray trajectory* shows evolution of the cost of resistance when the resistant strain is evolving alone in 107 × 107 lattice (the lattice size was shrunk so that the average density of resistant cells was roughly the same between simulations). In the case shown (and for several other simulations at a variety of lattice sizes), the solitary resistant strain immediately evolves to minimize its cost. Thus, the presence of producer and sensitive strains in a spatially structured habitat selects for competitive restraint in the resistant strain

By liberating the enemy of their enemy, these costly lineages liberate themselves (Tainaka 1993, 1995; Frean and Abraham 2001; Johnson and Seinen 2002). This phenomenon has been dubbed "survival of the weakest" (Frean and Abraham 2001).

6.5 Dynamics with many Strains: Universal Chemical Warfare

Naturally occurring microbial populations contain several different bacteriocinogenic strains (Gordon et al. 1998; Riley and Gordon 1999; Riley and Wertz 2002a, 2002b). Each distinct producer can be sensitive to the toxin produced by a different producer within its own (or closely related) species. In addition, resistance (sometimes to multiple toxins) can be generated through mutation and is very common in natural populations (Feldgarden and Riley 1998, 1999). What are the dynamical consequences of many interlacing games of rock–paper–scissors being played out simultaneously? How is such diversity maintained? It turns out that by inspecting such convoluted microbial chemical warfare, we gain some insight into mechanisms maintaining diversity (Lenski and Riley 2002).

There have been a few models that have considered multiple bacteriocin producers with cross-killing abilities. Pagie and Hogeweg (1999) model multiple producers within a lattice-based simulation framework. They find that within a spatially structured system, multiple producers can stably coexist. Further, the type of coexistence depends on the cost of resistance against toxins. If this cost is low, then the community enters into a "hyperimmunity" mode where most cells will be resistant to many different toxins, but few cells will produce very many toxins. However, if this cost is high, the community displays a "multitoxicity" mode, where cells are resistant to fewer toxins and tend to individually produce more toxins. Interestingly, the shift from "hyperimmunity" to "multitoxicity" is rather abrupt as the cost of resistance increases. Czárán et al. (2002) build on this earlier model of multiple producers, incorporating horizontal gene transfer and recombination between strains. Under many circumstances, they find that their lattice of multiple producers transitions through a "multitoxicity" mode and settles into a "hyperimmunity" mode.

Spatial structure is an important ingredient in these models – diversity drops dramatically in a well-mixed environment (Pagie and Hogeweg 1999). This role of spatial structure has been validated in a few experimental studies of multiple producers (Tait and Sutherland 2002; Massey et al. 2004). So far, most experimental work on multi-producer communities has been limited to pairs of interacting strains. It will be interesting to follow experimentally the dynamics of larger numbers of microbial players in order to see how well the predictions of the simulation models play out.

The study of communities with multiple producers will be especially exciting in light of recent observations suggesting that bacteriocins excreted by one producer can act as inducers of bacteriocins in other producers (Kuipers et al. 1995; Kleerebezem et al. 1997, 2004; Tait and Sutherland 2002; Gillor and Riley, unpublished data). Interestingly, such cross-induction can reintroduce bistability into the spatial dynamics of a two-strain system (Gillor et al., unpublished data). Specifically, cells of a rare invading producer (call the invading strain **A**) will induce neighboring resident producer cells (call the resident strain **B**) to produce toxin, which in turn will further induce the invader. This local escalation of chemical warfare can favor the common producer strain, as it effectively "surrounds" the invading strain with its toxin. This means that a population of producer strain **A** can exclude invading strain **B**, and a population of producer strain **B** can exclude invading strain **A**. It will be intriguing to see if this potential return to bistability occurs in spatially structured multiple-producer laboratory communities.

6.6 Discussion

Lewontin (1982, 1983) has suggested that the metaphor of adaptation (in which organisms that best fit preexistent niches are selected) should be replaced with a metaphor of construction. Lewontin's idea is that organisms, through their physiology, behavior and development, alter their world and thus influence the very form of their niche. That is to say, niches are not simply "out there" waiting to be filled, but rather are (at least partially) made via the effects organisms have on their abiotic and biotic surroundings. In Lewontin's view, the organism becomes a co-author in its own evolution and ecology. This process has been labeled niche construction (Odling-Smee et al. 1996, 2003; Laland et al. 1999, 2000), or alternatively ecosystem engineering (Jones et al. 1994, 1997). The production of bacteriocins within microbial communities is a potent form of niche construction – a producing cell alters the toxin concentration of its surroundings, shifting strain composition toward immune and resistant types.

Indeed, this toxic niche construction is one way to form a non-transitive competitive dynamic. Specifically, with regards to growth rate, the strain on the bottom of the totem pole (the producer) kills the strain at the top (the sensitive), thus creating a loop in the competitive interactions. Such non-transitivity has been found in other systems as well, including side-blotched lizards (Sinervo and Lively 1996), sessile marine invertebrates (Buss and Jackson 1979), and yeast (Paquin and Adams 1983). Theoretical work on non-hierarchically organized communities has shown that such interactions can promote the maintenance of biodiversity (Huisman and Weissing 1999; Huisman et al. 2001). Non-transitivity may be an important ingredient in the persistence of diverse bacteriocin communities, but it seems to require a partner to get the job done. This partner is population structure.

Because niche construction is ultimately frequency-dependent, toxin producers competing with sensitive cells in a well-mixed environment face a dynamical hump to get over (Adams et al. 1979; Chao and Levin 1981; Levin 1988; Durrett and Levin 1997; Iwasa et al. 1998). In unstructured habitats, the signature of this hump is present in the resistant–producer–sensitive community, changing the dynamics from a straightforward "rock–paper–scissors" to a "one-winner" outcome (compare Fig. 6.2a, b to c, d). Population structure (e.g., spatial structure) can effectively eliminate the hump (Chao and Levin 1981; Durrett and Levin 1997; Iwasa et al. 1998) and restore the game of rock–paper–scissors. In this spatial game, players stably chase each other around a structured arena as clumps, with balanced gains and losses occurring at the boundaries (Durrett and Levin 1997; Kerr et al. 2002). Kirkup and Riley (2004) demonstrated that the same non-transitive dynamic occurs in the mouse alimentary tract, a spatially structured habitat. In addition, spatial structure is an important ingredient in the coexistence of multiple-producer strains (Pagie and Hogeweg 1999; Czárán et al. 2002).

It is worthwhile to highlight the nature of the explanations of biodiversity maintenance offered by the above models. While biodiversity in the system can result from exogenous heterogeneity in the underlying substrate (Frank 1994), many of these models describe diversity resulting from *endogenous* processes. That is, diversity is a product of the way non-transitive interactions play out in a spatially structured world. In this sense, diversity "flows from within" the system. Part of the recent interest in spatial ecology (Durrett and Levin 1994a, 1994b; May 1999; Bolker et al. 2003) derives from an interest in understanding how global patterns result from local processes (Hassell et al. 1994; May 1999; Wootton 2001). This idea of system self-organization is the natural outgrowth of localized niche construction, where the effects organisms have within neighborhoods scale up to influence the form of the entire community.

Models have been indispensable in the study of bacteriocin community self-organization. Part of this success has depended on the sustained interaction between those exploring theoretical models and those experimenting with model communities in the laboratory. For instance, Chao and Levin (1981) described frequency-dependence in sensitive–producer communities of well-mixed *E. coli*, and then Levin (1988) analytically demonstrated the bistability. As another example, Chao and Levin (1981) demonstrated that the spatial structure afforded by soft agar poured in a Petri dish could eradicate the frequency-dependence and then Durrett and Levin (1997), using cellular automata, confirmed these empirical observations (see also Iwasa et al. 1998). As yet another example, Durrett and Levin (1997), using lattice-based models, predicted that spatial structure would be required for long-term coexistence of the resistant–producer–sensitive community, and this was empirically confirmed 5 years later by Kerr et al. (2002). There has been mutual benefit by maintaining an active dialogue between theoretical and empirical work.

And such dialogue will certainly facilitate future understanding of these communities. There are several questions ripe for exploration. What are the evolutionary dynamics in resistant–producer–sensitive communities? Will we actually observe a form of "survival of the weakest" in laboratory communities? What are the ecological and evolutionary dynamics of communities with multiple bacteriocin producers? What are the dynamics of diverse bacteriocin communities in natural settings? What effects will cross-induction (another form of niche construction) play in these dynamics? Models will most certainly continue to play an important role in exploring such issues, and through the study of model systems, the next set of questions will begin to emerge.

Acknowledgements. I thank Milind Chavan, Carla Goldstone, Federico Prado, Peg Riley and Karen Walag for many useful comments on previous versions of this chapter.

Appendix

Sensitivity is an ESS in the Well-Mixed RPS Game

Consider the following set of differential equations describing the dynamics of sensitive, producer and resistant strains (see Durrett and Levin 1997, and Box 1):

$$\frac{ds}{dt} = \beta_s(1-s-p-r)s - (\delta_s + \gamma p)s, \tag{A.1}$$

$$\frac{dp}{dt} = \beta_p(1-s-p-r)p - \delta_p p, \tag{A.2}$$

$$\frac{dr}{dt} = \beta_r(1-s-p-r)r - \delta_r r. \tag{A.3}$$

This system has an equilibrium at $(s,p,r) = ((\beta_s - \delta_s)/\beta_s, 0, 0) = (\hat{s}, 0, 0)$ where only toxin-sensitive cells exist. Consider a perturbation to this equilibrium, $(\hat{s} + \varepsilon_s, \varepsilon_p, \varepsilon_r)$, where all ε values are very small. The dynamics of the perturbations are given by:

$$\frac{d\varepsilon_s}{dt} = (\beta_s(1-2\hat{s}) - \delta_s)\varepsilon_s - ((\beta_s + \gamma)\hat{s})\varepsilon_p - (\beta_s \hat{s})\varepsilon_r - \beta_s \varepsilon_s(\varepsilon_s + \varepsilon_p + \varepsilon_r) - \gamma \varepsilon_s \varepsilon_p, \tag{A.4}$$

$$\frac{d\varepsilon_p}{dt} = (\beta_p(1-\hat{s}) - \delta_p)\varepsilon_p - \beta_p \varepsilon_p(\varepsilon_s + \varepsilon_p + \varepsilon_r), \tag{A.5}$$

$$\frac{d\varepsilon_r}{dt} = (\beta_r(1-\hat{s}) - \delta_r)\varepsilon_r - \beta_r \varepsilon_r(\varepsilon_s + \varepsilon_p + \varepsilon_r). \tag{A.6}$$

Linearizing the system about $(\varepsilon_s, \varepsilon_p, \varepsilon_r) = (0,0,0)$, we have

$$\dot{\vec{\varepsilon}} = J\vec{\varepsilon}, \tag{A.7}$$

where

$$\vec{\varepsilon} = \begin{bmatrix} \varepsilon_s \\ \varepsilon_p \\ \varepsilon_r \end{bmatrix}, \tag{A.8}$$

and the Jacobian is

$$\mathbf{J} = \begin{bmatrix} \beta_s(1-2\hat{s}) - \delta_s & -(\beta_s + \gamma)\hat{s} & -\beta_s \hat{s} \\ 0 & \beta_p(1-\hat{s}) - \delta_p & 0 \\ 0 & 0 & \beta_r(1-\hat{s}) - \delta_r \end{bmatrix}. \tag{A.9}$$

The eigenvalues of $\mathbf{J}$ give the local stability of the equilibrium $(\hat{s},0,0)$. Since $\mathbf{J}$ is a triangular matrix, the eigenvalues line the diagonal. Because we assume

$$\beta_i > \delta_i \text{ for all } i \in \{s,p,r\}, \tag{A.10}$$

and

$$\frac{\delta_s}{\beta_s} < \frac{\delta_r}{\beta_r} < \frac{\delta_p}{\beta_p} < \frac{\delta_s + \gamma}{\beta_s + \gamma}, \tag{A.11}$$

all of the eigenvalues are negative, which means the equilibrium $(\hat{s},0,0)$ is locally stable and thus toxin sensitivity is an evolutionarily stable strategy (an ESS). The other fixation equilibria,

$$e_2 = \left(0, \frac{\beta_p - \delta_p}{\beta_p}, 0\right), \tag{A.12}$$

$$e_3 = \left(0, 0, \frac{\beta_r - \delta_r}{\beta_r}\right), \tag{A.13}$$

are locally unstable (this can be shown using linear stability analysis as well). Lastly, under assumption (A.11), there is another equilibrium:

$$e_4 = \left(1 + \frac{\delta_s}{\gamma} - \left(1 + \frac{\beta_s}{\gamma}\right)\frac{\delta_p}{\beta_p}, \frac{1}{\gamma}\left(\frac{\delta_p}{\beta_p}\beta_s - \delta_s\right), 0\right), \tag{A.14}$$

which also is unstable. Thus, sensitivity to the toxin is the only ESS in this system. Indeed, from nearly any starting point, the sensitive strain will displace the other two strains (Nakamaru and Iwasa 2000).

References

Adams J, Kinney T, Thompson S, Rubin L, Helling RB (1979) Frequency-dependent selection for plasmid-containing cells of *Escherichia coli*. Genetics 91:627–637

Bolker BM, Pacala SW, Neuhauser C (2003) Spatial dynamics in model plant communities: what do we really know? Am Naturalist 162:135–148
Buss LW, Jackson JBC (1979) Competitive networks – non-transitive competitive relationships in cryptic coral-reef environments. Am Naturalist 113:223–234
Chao L, Levin BR (1981) Structured habitats and the evolution of anticompetitor toxins in bacteria. Proc Natl Acad Sci USA Biol Sci 78:6324–6328
Czárán TL, Hoekstra RF (2003) Killer-sensitive coexistence in metapopulations of microorganisms. Proc R Soc Lond Series B-Biol Sci 270:1373–1378
Czárán TL, Hoekstra RF, Pagie L (2002) Chemical warfare between microbes promotes biodiversity. Proc Natl Acad Sci USA 99:786–790
Durrett R, Levin S (1994a) The importance of being discrete (and spatial). Theor Popul Biol 46:363–394
Durrett R, Levin SA (1994b) Stochastic spatial models – a users guide to ecological applications. Philos Trans R Soc Lond Series B-Biol Sci 343:329–350
Durrett R, Levin S (1997) Allelopathy in spatially distributed populations. J Theor Biol 185:165–171
Dykes GA, Hastings JW (1997) Selection and fitness in bacteriocin-producing bacteria. Proc R Soc Lond Series B-Biol Sci 264:683–687
Feldgarden M, Riley MA (1998) High levels of colicin resistance in *Escherichia coli*. Evolution 52:1270–1276
Feldgarden M, Riley MA (1999) The phenotypic and fitness effects of colicin resistance in *Escherichia coli* K-12. Evolution 53:1019–1027
Frank SA (1994) Spatial polymorphism of bacteriocins and other allelopathic traits. Evol Ecol 8:369–386
Frean M, Abraham ER (2001) Rock-scissors-paper and the survival of the weakest. Proc R Soc Lond Series B-Biol Sci 268:1323–1327
Gilpin ME (1975) Limit cycles in competition communities. Am Naturalist 109:51–60
Gordon DM, Riley MA (1999) A theoretical and empirical investigation of the invasion dynamics of colicinogeny. Microbiology-SGM 145:655–661
Gordon DM, Riley MA, Pinou T (1998) Temporal changes in the frequency of colicinogeny in *Escherichia coli* from house mice. Microbiology-UK 144:2233–2240
Hassell MP, Comins HN, May RM (1994) Species coexistence and self-organizing spatial dynamics. Nature 370:290–292
Huisman J, Weissing FJ (1999) Biodiversity of plankton by species oscillations and chaos. Nature 402:407–410
Huisman J, Johansson AM, Folmer EO, Weissing FJ (2001) Towards a solution of the plankton paradox: the importance of physiology and life history. Ecol Lett 4:408–411
Iwasa Y, Nakamaru M, Levin SA (1998) Allelopathy of bacteria in a lattice population: competition between colicin-sensitive and colicin-producing strains. Evol Ecol 12:785–802
James R, Kleanthous C, Moore GR (1996) The biology of E colicins: paradigms and paradoxes. Microbiology-UK 142:1569–1580
Johnson CR, Seinen I (2002) Selection for restraint in competitive ability in spatial competition systems. Proc R Soc Lond Series B-Biol Sci 269:655–663
Jones CG, Lawton JH, Shachak M (1994) Organisms as ecosystem engineers. Oikos 69:373–386
Jones CG, Lawton JH, Shachak M (1997) Positive and negative effects of organisms as physical ecosystem engineers. Ecology 78:1946–1957
Kerr B, Riley MA, Feldman MW, Bohannan BJM (2002) Local dispersal promotes biodiversity in a real-life game of rock-paper-scissors. Nature 418:171–174
Kirkup BC, Riley MA (2004) Antibiotic-mediated antagonism leads to a bacterial game of rock-paper-scissors in vivo. Nature 428:412–414
Kleerebezem M, Quadri LEN, Kuipers OP, de Vos WM (1997) Quorum sensing by peptide pheromones and two-component signal-transduction systems in Gram-positive bacteria. Mol Microbiol 24:895–904

Kleerebezem M, Bongers R, Rutten G, de Vos WM, Kuipers OP (2004) Autoregulation of subtilin biosynthesis in *Bacillus subtilis*: the role of the spa-box in subtilin-responsive promoters. Peptides 25:1415–1424

Kuipers OP, Beerthuyzen MM, Deruyter P, Luesink EJ, de Vos WM (1995) Autoregulation of nisin biosynthesis in *Lactococcus lactis* by signal-transduction. J Biol Chem 270:27299–27304

Laland KN, Odling-Smee FJ, Feldman MW (1999) Evolutionary consequences of niche construction and their implications for ecology. Proc Natl Acad Sci USA 96:10242–10247

Laland KN, Odling-Smee FJ, Feldman MW (2000) Niche construction, biological evolution and cultural change. Behav Brain Sci 23:131–175

Lenski RE, Riley MA (2002) Chemical warfare from an ecological perspective. Proc Natl Acad Sci USA 99:556–558

Levin BR (1988) Frequency-dependent selection in bacterial-populations. Philos Trans R Soc Lond Series B-Biol Sci 319:459–472

Lewontin RC (1982) Organism and environment. In: Learning, development, and culture. MIT Press, Cambridge, MA

Lewontin RC (1983) Gene, organism, and environment. In: Bendall DS (ed) Evolution from molecules to men. Cambridge University Press, Cambridge

Massey RC, Buckling A, Ffrench-Constant R (2004) Interference competition and parasite virulence. Proc R Soc Lond Series B-Biol Sci 271:785–788

May R (1999) Unanswered questions in ecology. Philos Trans R Soc Lond Series B-Biol Sci 354:1951–1959

Nakamaru M, Iwasa Y (2000) Competition by allelopathy proceeds in traveling waves: colicin-immune strain aids colicin-sensitive strain. Theor Popul Biol 57:131–144

Odling-Smee FJ, Laland KN, Feldman MW (1996) Niche construction. Am Naturalist 147:641–648

Odling-Smee FJ, Laland KN, Feldman MW (2003) Niche construction: the neglected process in evolution. Princeton University Press, Princeton, NJ

Pagie L, Hogeweg P (1999) Colicin diversity: a result of eco-evolutionary dynamics. J Theor Biol 196:251–261

Paquin CE, Adams J (1983) Relative fitness can decrease in evolving asexual populations of *S. cerevisiae*. Nature 306:368–371

Riley MA (1998) Molecular mechanisms of bacteriocin evolution. Annu Rev Genet 32:255–278

Riley MA, Gordon DM (1996) The ecology and evolution of bacteriocins. J Indus Microbiol 17:151–158

Riley MA, Gordon DM (1999) The ecological role of bacteriocins in bacterial competition. Trends Microbiol 7:129–133

Riley MA, Wertz JE (2002a) Bacteriocin diversity: ecological and evolutionary perspectives. Biochimie 84:357–364

Riley MA, Wertz JE (2002b) Bacteriocins: evolution, ecology, and application. Annu Rev Microbiol 56:117–137

Sinervo B, Lively CM (1996) The rock-paper-scissors game and the evolution of alternative male strategies. Nature 380:240–243

Tainaka K (1993) Paradoxical effect in a 3-candidate voter model. Phys Lett A 176:303–306

Tainaka K (1995) Indirect effect in cyclic voter models. Phys Lett A 207:53–57

Tait K, Sutherland IW (2002) Antagonistic interactions amongst bacteriocin-producing enteric bacteria in dual species biofilms. J Appl Microbiol 93:345–352

Tan Y, Riley MA (1996) Rapid invasion by colicinogenic *Escherichia coli* with novel immunity functions. Microbiology-UK 142:2175–2180

Wiener P (2000) Antibiotic production in a spatially structured environment. Ecol Lett 3:122–130

Wootton JT (2001) Local interactions predict large-scale pattern in empirically derived cellular automata. Nature 413:841–844

7 Bacteriocins' Role in Bacterial Communication

OSNAT GILLOR

Summary

Intercellular communication and multicellular coordination are now known to be widespread among prokaryotes and to affect multiple phenotypes. Bacterial cell–cell communication involves sophisticated signal-transduction networks aimed at integrating intercellular signals with other information for making decisions about gene expression and cellular differentiation. Many different classes of signaling molecules have been identified in both Gram-positive and Gram-negative bacterial species. One of the more surprising groups of intercellular signaling molecules was recently identified in the bacteriocins. Along with their role as antibacterial toxins, bacteriocins were found to act as signaling and coordinating agents necessary for invading, establishing and competing in natural environments. In this chapter, I will review the newly discovered role bacteriocins play in regulating microbial group behavior, such as quorum sensing and biofilm formation.

7.1 Introduction

Cell–cell communication is a fundamental activity performed by most types of cells. Previously, scientists held the view that bacterial cells behaved as self-sufficient individuals, unable to organize themselves into groups or communicate. Bacteria were considered as nothing more than a mass of individuals scavenging for nutrients and multiplying independently. However, recent studies have shown that bacteria are capable of coordinated activities. At first, the idea that bacteria could function as groups, and that individuals within the group could respond to the group as a whole seemed almost ludicrous. These sorts of interactions were attributed only to more "highly evolved" organisms. Nevertheless, nowadays it is generally accepted that

---

Department of Environmental Hydrology & Microbiology, Zuckerberg Institute for Water Research, Jacob Blaustein Institute for Desert Research, Ben Gurion University of the Negev, Sde Boqer Campus 84990, Israel, e-mail: gilloro@bgu.ac.il

bacteria produce, and respond as groups to signals and that this interaction can lead to the coordination of group bacterial activities (Shapiro 1998; Greenberg 2003a, 2003b). The change in thinking is due mainly to the discovery of quorum sensing signal molecules, which are used throughout the eubacterial kingdom to regulate the expression of a wide variety of phenotypes (Fuqua et al. 1994). It was further supported by the discovery of auto-aggregation of chemotactic bacteria, and coordinated behaviors in complex colony morphogenesis (Armitage et al. 2003; Smith et al. 2004).

Quorum sensing is used to describe a collection of molecular mechanisms that are employed by bacteria to monitor density (Bassler 1999; Miller and Bassler 2001). Bacterial intercellular communication is based on the detection of diffusible signal molecules. Bacteria use a wide variety of signaling molecules, signal-detection systems and signal-transduction mechanisms to convert the information contained in the signal into changes in gene regulation. Quorum sensing allows populations of bacteria to collectively regulate gene expression and synchronize group behavior. This synchronization is generally timed to coincide with attaining a high population density. In Gram-negative bacteria, the signaling molecules are often acylated homoserine lactones (Fuqua et al. 2001; Whitehead et al. 2001); in Gram-positive bacteria, signaling molecules are often peptides (Kleerebezem et al. 1997b).

Bacterial quorum sensing enhances access to nutrients or environmental niches, enables bacteria to mount defensive response against competing organisms, and optimizes the ability of the cell to differentiate into morphological forms better adapted to survival in a hostile environment (Miller and Bassler 2001). The current dogma is that a population of cells occupies a closed ecological niche synthesizing constitutive low-level signal, and that its concentration rises in synchrony with the increase in the cellular population (Smith et al. 2004). Systems have apparently evolved so that the signal reaches a concentration sufficient for biological activity at a cell density that is appropriate for the induced activity to occur efficiently (Swift et al. 2001). Bacteria use quorum sensing systems to regulate several physiological properties, including the ability to incorporate foreign DNA (Havarstein et al. 1995; Li et al. 2001), acid-tolerant response (Li et al. 2002), virulence regulation (Bauer and Robinson 2002), and biofilm formation (Li et al. 2001).

Biofilms are dense aggregates of surface-adherent microorganisms embedded in an exopolysaccharide matrix. The study of bacteria residing in biofilms as an interactive community, rather than free-living planktonic cells, has recently gained a great deal of attention (Watnick and Kolter 2000; Greenberg 2003a, 2003b). Genetic studies of single-species biofilms have shown that they form in multiple steps (Pratt and Kolter 1998, 1999; Watnick and Kolter 1999), require intercellular signaling (Davies et al. 1998), and demonstrate a profile of gene transcription that is distinct from that of planktonic cells (Prigent-Combaret et al. 1999). In natural environments, the biofilm is almost invariably composed of mixed microbial cultures, which undergo a wide range of physiological and morphological adaptations in response to the changing environment. Different gradients of chemicals, nutrients and oxygen create microenvironments within the biofilm to which the bacteria must adapt to survive (Watnick and Kolter 2000).

In competition with other species for the same nutrients or niche, most bacterial species produce bacteriocins, ribosomally synthesized peptides or proteins with antibacterial activity (Riley and Wertz 2002). Although bacteriocins are not required for growth, they may help microorganisms that produce them to compete for resources in their environment (Riley and Gordon 1999; Kerr et al. 2002). In this chapter, I will review the role of bacteriocins in mediating intercellular communication and multicellular behavior. I will show how, along with their role as antibacterial toxins, bacteriocins can act as agents in cell–cell communication and coordination necessary for invading, establishing and competing in natural environments.

7.2 Bacteriocin-Mediated Intercellular Communication

Intercellular communication is the basis of coordinated multicellular function. The molecular basis of intercellular coordination is being clarified in many bacterial species, and homologies have been discovered between intercellular and unicellular regulatory circuits (O'Toole and Kolter 1998a, 1998b; Pratt and Kolter 1998; Watnick and Kolter 1999). In Gram-negative bacteria, acyl-homoserine lactone molecules often serve as the signals, and there is a well-documented mechanism by which the signals are recognized and converted to a functional response by the organisms. There is also evidence for lipid-, oligopeptide- and amino acid-based signaling (Fuqua et al. 2001; Whitehead et al. 2001; Smith et al. 2004). I will describe cell–cell signaling systems in Gram-positive bacteria, as these organisms use peptides or modified peptides as signals (Dunny and Leonard 1997), which correlates well with the bacteriocins produced by this group of bacteria (unlike the bacteriocins produced by Gram-negative bacteria; Riley and Wertz 2002).

In Gram-positive bacteria, quorum sensing systems generally consist of three components – a peptide pheromone, which acts as the signal peptide, and a two-component regulatory system (also called two-component signal-transduction system) that has a membrane-bound histidine kinase sensor and an intracellular response regulator. The secreted pheromone binds to the histidine kinase, resulting in autophospharylation, the phosphoryl group is then transferred to the response regulator, which binds to the regulated promoters and activates them (Dunny and Leonard 1997; Kleerebezem et al. 1997b; Smith et al. 2004). Quorum sensing in Gram-positive bacteria has been found to regulate a number of physiological activities, including competence development in *Streptococcccus gordonii*, *S. pneumoniae*, and *S. mutans* (Cvitkovitch 2001; Cvitkovitch et al. 2003), sporulation in *Bacillus subtilus* (Lazazzera 2000), induction of virulence factors in *Staphylococcus aureus*

(Otto 2001), and bacteriocin biosynthesis in *Lactococcus lactis* (Kuipers et al. 1995). In the following section, I will describe the autoregulation of two bacteriocin groups both produced by Gram-positive bacteria.

7.2.1 Autoregulation of Class I Bacteriocins

The class I bacteriocins include the lantibiotics, which are heat-stable and contain post-translationally modified amino acids (McAuliffe et al. 2001). The best-studied lantibiotic is by far nisin, produced by *Lactococcus lactis*. Nisin serves as a model system for investigations of post-translationally modified bacteriocin structure/function relationships, genetic organization, and biochemical properties (Carr et al. 2002). It was further suggested that this peptide not only acts as a toxin against other microorganisms, but can also serve as a quorum sensing signal for the expression of biosynthesis and immunity genes in the nisin-producing cells (Kuipers et al. 1995; Kleerebezem et al. 1997a).

Nisin production is encoded by the chromosomally located gene cluster nisABTCIPRKFEG. The *nisA* gene encodes the pre-peptide, whereas *nisB*, *nisC*, *nisT*, and *nisP* genes encode enzymes that are involved in a series of post-translational modifications, export, and proteolytic processing steps required for producing the mature lantibiotic. The *nisI* and *nisFEG* are involved in the producer cell immunity against nisin, while *nisR* (response regulator) and *nisK* (sensor kinase) genes serve for two-component signal transduction (Eichenbaum et al. 1998; Twomey et al. 2002).

A deletion in *nisA* leads to the abolition of transcription of the gene and loss of nisin production. Addition of sublethal quantities of nisin to the medium restored the transcription of *nisA*, suggesting that the product of the pathway in this model serves as an extracellular signal for further transcriptional activation. Activation of NisK (the sensor kinase) by nisin leads to autophosphorylation of the protein with subsequent phosphotransfer to NisR (the response regulator), which acts on the nisin gene cluster. These data suggest that nisin can act as an extracellular pheromone involved in the regulation of its own synthesis (Kuipers et al. 1995).

It has been shown that gene clusters encoding the production of bacteriocins isolated from *S. salivarius*, *S. pyogenes* (Cvitkovitch et al. 2003) and *Bacillus subtillis* (Kleerebezem et al. 1997b) all contain genes encoding the two-component (sensor–regulator) system, suggesting that their bacteriocin productions are also autoregulated by a mechanism similar to that of nisin.

7.2.2 Quorum Sensing Regulation of Class II Bacteriocins

The class II bacteriocins are heat-stable and do not contain modified amino acids. In fact, unlike group I, the quorum sensing peptide of group II bacteriocins is not a bacteriocin but an inactive peptide that in structure is closely related to the bacteriocin, and referred to as the inducing peptide. It was also found that one or more operons (often linked) that encode bacteriocin biosynthesis and immunity are co-regulated with the induction factor production, i.e., they are transcriptionally activated in response to the inducing peptide-mediated signal-transduction pathway. The pre-inducing peptide is cleaved during its transport to the outside of the cell, resulting in the formation of the inducing peptide, which is then sensed by the two-component regulatory system (the sensor kinase and the response regulator). The signal transducted by the inducing peptide is subsequently transferred to the histidine protein kinase, which in turn transduces the signal of the response regulator, a DNA-binding protein that activates the genes responsible for bacteriocin synthesis (Nes et al. 1996; Nes and Holo 2000; Eijsink et al. 2002), in a manner similar to that described for class I bacteriocins.

7.3 Bacteriocin-Coordinated Multicellular Communication

Most microbiologists forget that the well-aerated planktonic culture is largely a laboratory construct, and that in nature most bacteria proliferate and survive attached to surfaces. Surface cultures are usually an aggregation of multiple bacterial species that differentiate biochemically and morphologically. The most widespread structures of multicellular prokaryotes in nature are the biofilms (Watnick and Kolter 2000).

Biofilms are communities of microorganisms enclosed in distinct three-dimensional structures with fluid channels for transport of substrate, waste products, and signal molecules. The matrix that holds the biofilm together is a mixture of polysaccharides, proteins and DNA secreted by the cells (Scheie and Petersen 2004). Biofilms consist of single or multiple microbial species, and are found on a variety of biotic and abiotic surfaces (Shapiro 1998). The formation of a biofilm is dependent on quorum sensing. It is a stepwise process, which involves adhesion of planktonic microorganisms to a surface, colonization and co-adhesion, growth and maturation, and finally detachment of some of the microorganisms. Evidence is emerging that gene expression required during the various stages of development is coordinated between the different species inhabiting the biofilm. Various signal-transduction systems induce cascades of reactions leading to the induction and inhibition of gene transcription in accordance to the biofilm state (Pratt and Kolter 1999).

Bacterial products able to diffuse from one cell to another generally carry out communication between bacteria. This is probably not effective between planktonic bacteria in aquatic environments, because the signaling molecules are likely to be too diluted, with a very small probability of reaching their target. However, this method of signaling seems ideally suited for bacteria in a biofilm (Shapiro 1998; Watnick and Kolter 2000). Although little is known of the role of intercellular signaling in multispecies biofilms, these signals should be particularly important in environments where surfaces are heavily colonized, and competition for available nutrients and space is strong. Cell–cell communication in biofilms might include bacterial metabolites, genetic material or secreted peptides, and may result in the distribution of specific species, modification of protein expression or introduction of new genetic traits into neighboring cells (Greenberg 2003b; Cvitkovitch et al. 2003; Scheie and Petersen 2004; Smith et al. 2004).

Bacteriocins could be one example for intercellular signals used within a biofilm. In fact, mathematical models predict that bacteriocin production would be most advantageous in a spatially structured environment such as a biofilm, suggesting that these secreted proteins may have evolved specifically for the biofilm environment (Frank 1994; Durrett and Levin 1997). Here, I describe the role bacteriocins play in two extensively studied biofilm hosts – the oral cavity and the gastrointestinal (GI) tract.

7.3.1 Oral Biofilms

The human oral cavity is a complex ecosystem that supports an extremely diverse microflora of nearly 500 species of microorganisms (Kroes et al. 1999). Numerous physical and nutritional interactions between oral bacteria contribute to this complex biofilm community (Scheie and Petersen 2004). The spatial organization of the species within each biofilm is unique, although the most frequently isolated species make a major contribution to each community covering the teeth surface. Streptococci and actinomyces are the major initial colonizers of the tooth surface, and the interactions between these and their substrata help establish the early biofilm community (Palmer et al. 2003). Fusobacteria play a central role as physical bridges that mediate co-aggregation of cells, and as physiological bridges that promote anaerobic microenvironments, which protect the co-aggregating of strict anaerobes in an aerobic atmosphere (Kolenbrander et al. 1989).

Communication among the microorganisms retaining the biofilm is essential for initial colonization and subsequent maturation on the enamel surfaces of the teeth, without which some species would be swallowed with the saliva. Through retention, the oral bacteria can form organized multispecies communities commonly referred to as dental plaque (Kolenbrander et al. 2002).

Transfer of genes by competence-inducing pathways is one of the most studied forms of communication by oral bacteria. The competent state permits the binding, uptake and integration of extracellular DNA to occur. It is thought that this system influences the ability to adapt to the environment by promoting the acquisition of new genetic traits from other bacteria. For example, sensing the bacteriocin salivaricin A by its producer *Streptococcus salivarius* regulates a two-component system comprised of the histidine kinase SalK and the cognate response regulator SalR. Interestingly, *S. salivarius* can sense a salivaricin A homologue produced by *S. pyogenes*, suggesting a mechanism of interspecies communication (Upton et al. 2001). Another example would be the disruption of *S. gordonii* genes involved in the early stages of competence development for transformation, resulting in mutants deficient in their ability to produce the two bacteriocins STH1 and STH2. This link suggests that killing microorganisms by bacteriocins could serve dual functions – to liberate DNA for uptake by the competent bacteriocin producers, and to prevent colonization by invading microorganisms (Yother et al. 2002).

Mutacin IV, a bacteriocin produced by *S. mutans*, abolishes the growth of the closely related species *S. gordonii*, and has been linked to the competence development system in the producer strain (Qi et al. 2001). Culture assays of the two strains demonstrated that increased transformation in *S. murtans* by plasmid DNA, originally harbored in *S. gordonii*, was dependent on the presence of mutacin IV genes. Upon addition of partially purified mutacin IV, there was an increased DNA release by *S. gordonii*. These findings suggest a coordinated production of the bacteriocin and development of competence is a possible mechanism for DNA uptake in a multispecies microbial community (Kreth et al. 2005). An interesting twist to this story was recently reported: a protease produced by *S. gordonii* Challis was found to significantly reduce bacteriocin production by *S. mutans* in a biofilm, but not in broth. It was also shown that interactions with other oral streptococci in a biofilm, but not in broth, inhibited the bacteriocin production of *S. mutans* (Wang and Kuramitsu 2005). This finding suggests a form of defense has evolved in the target cells.

Fusobacterium nucleatum plays a key role in oral biofilm formation by promoting the co-aggregation of other bacteria, especially Gram-negative bacilli, which are the predominant microorganisms when an oral plaque is completely settled. It was recently demonstrated that a bacteriocin produced by *F. nucleatum* inhibits mainly lactobacilli, suggesting that this bacterium is not only important for its co-aggregation capacity, but it also intervenes by regulating other species numbers inside the plaque (Testa et al. 2003).

The studies described above suggest that determinants of population dynamics encompass subtle combinations of complex environmental sensing systems that are not limited to cell density, bacteriocin production, and competence stimulation.

7.3.2 Gastrointestinal Biofilms

Bacteria in the GI tract comprise a complex, multispecies community, its members interacting with each other as well as with their animal host. In a mature adult cultivation, the majority are anaerobic species detected together with some aerobic and facultative species. Distinct differences were reported between the bacterial composition of the cecum, colon, and feces as well as between the luminal and mucosal layers (Anderson 2003). Many GI bacteria readily adhere to the GI tract epithelial layer, as the first step toward biofilm formation (Probert and Gibson 2002). Biofilms in the GI tract would invariably be composed of large multispecies communities, although some biofilms may be a more favorable habitat for certain bacterial species. A number of factors can influence the GI bacterial composition, and as a result, the composition and proportion of member species within the GI tract, at any given niche, are constantly changing. A biofilm community can more readily supply an intestinal species with essentials than if it had remained in the planktonic phase. However, cooperation may not always be the goal, and antagonistic interactions may play an important role in the development of microbial community in the gut (Anderson 2003).

Bacteriocins are produced by all major groups of Archeae and Bacteria. Indeed, 35% of *Escherichia coli* isolated from the human GI tract produce bacteriocins (Riley and Gordon 1992). Given their abundance in the intestinal tracts, it is likely that they play a substantial role in colon biofilm interactions. It was previously stated that in planktonic cultures bacteriocin-producing and -sensitive strains cannot coexist (Riley and Gordon 1999), but this might not be true of biofilms. Bacteriocin-sensitive *Enterobacter agglomerans* and *E. gergoviae*, together with a bacteriocin-producing *E. coli* strain, co-inhabited dual-species biofilms (at all possible combinations). The established dual biofilms suggest that bacteriocin-sensitive and -producing bacteria can coexist in the same biofilm, though the bacteriocin-producing organisms do have a competitive advantage (Tait and Sutherland 2002). Bacteriocin-producing *Bacteriodes* strains were also found to coexist in the colon with a larger population of bacteriocin-susceptible strains of *Bacteriodes* (Booth et al. 1977).

Likewise, in established biofilms some of the members secrete bacteriocin apparently in an attempt to prevent undesirable competitors from colonizing their biofilm. In a GI tract model, the bacteriocin-producing *Lactobacillus curvatus* provided protection against *E. coli* and *Listeria innocua* invasion. These data suggest that bacteriocin-producing lactobacilli established in the human colon prevent new strains from invading or maintaining stable populations (Ganzle et al. 1999).

7.4 Conclusions

Lately, we have witnessed a remarkable increase in research on prokaryotic cell–cell communication systems. These systems have been shown to play critical roles in controlling the expression of some of the most important biological functions of microorganisms. Bacteriocins, which are primarily identified as antibiotic proteins active against closely related species, have recently been considered as intercellular signaling molecules and are suggested to play a major role in regulating collective microbial behaviors. An increased understanding of a bacteriocin's role as a signaling peptide, together with its traditional role as a toxin, may lead to useful applications. These include an improved ability to identify novel targets for the development of new drugs, as well as the expression of useful metabolic functions for biotechnological applications.

References

Anderson KL (2003) The complex world of gastrointestinal bacteria. Can J Anim Sci 83:409–427

Armitage JP, Dorman CJ, Hellingwerf K, Schmitt R, Summers D, Holland B (2003) Thinking and decision making, bacterial style. In: Proc Conf Bacterial Neural Networks, 7–12 June 2002, Obernai, France. Mol Microbiol 47:583–593

Bassler BL (1999) How bacteria talk to each other: regulation of gene expression by quorum sensing. Curr Opin Microbiol 2:582–587

Bauer WD, Robinson JB (2002) Disruption of bacterial quorum sensing by other organisms. Curr Opin Biotechnol 13:234–237

Booth SJ, Johnson JL, Wilkins TD (1977) Bacteriocin production by strains of Bacteroides isolated from human feces and the role of these strains in the bacterial ecology of the colon. Antimicrob Agents Chemother 11:718–724

Carr FJ, Chill D, Maida N (2002) The lactic acid bacteria: a literature survey. Crit Rev Microbiol 28:281–370

Cvitkovitch DG (2001) Genetic competence and transformation in oral *Streptococci*. Crit Rev Oral Biol Med 12:217–243

Cvitkovitch DG, Li YH, Ellen RP (2003) Quorum sensing and biofilm formation in Streptococcal infections. J Clin Invest 112:1626–1632

Davies DG, Parsek MR, Pearson JP, Iglewski BH, Costerton JW, Greenberg EP (1998) The involvement of cell-to-cell signals in the development of a bacterial biofilm. Science 280:295–298

Dunny GM, Leonard BA (1997) Cell-cell communication in gram-positive bacteria. Annu Rev Microbiol 51:527–564

Durrett R, Levin S (1997) Allelopathy in spatially distributed populations. J Theor Biol 185:165–171

Eichenbaum Z, Federle MJ, Marra D, de Vos WM, Kuipers OP, Kleerebezem M, Scott JR (1998) Use of the lactococcal nisA promoter to regulate gene expression in gram-positive bacteria: comparison of induction level and promoter strength. Appl Environ Microbiol 64:2763–2769

Eijsink VG, Axelsson L, Diep DB, Havarstein LS, Holo H, Nes IF (2002) Production of class II bacteriocins by lactic acid bacteria; an example of biological warfare and communication. Antonie Van Leeuwenhoek 81:639–654

Frank SA (1994) Spatial polymorphism of bacteriocins and other allelopathic traits. Evol Ecol 8:369–386

Fuqua WC, Winans SC, Greenberg EP (1994) Quorum sensing in bacteria: the LuxR-LuxI family of cell density-responsive transcriptional regulators. J Bacteriol 176:269–275

Fuqua C, Parsek MR, Greenberg EP (2001) Regulation of gene expression by cell-to-cell communication: acyl-homoserine lactone quorum sensing. Annu Rev Genet 35:439–468

Ganzle MG, Hertel C, van der Vossen JM, Hammes WP (1999) Effect of bacteriocin-producing lactobacilli on the survival of *Escherichia coli* and *Listeria* in a dynamic model of the stomach and the small intestine. Int J Food Microbiol 48:21–35

Greenberg EP (2003a) Bacterial communication and group behavior. J Clin Invest 112:1288–1290

Greenberg EP (2003b) Bacterial communication: tiny teamwork. Nature 424:134
Havarstein LS, Coomaraswamy G, Morrison DA (1995) An unmodified heptadecapeptide pheromone induces competence for genetic transformation in *Streptococcus pneumoniae*. Proc Natl Acad Sci USA 92:11140–11144
Kerr B, Riley MA, Feldman MW, Bohannan BJ (2002) Local dispersal promotes biodiversity in a real-life game of rock-paper-scissors. Nature 418:171–174
Kleerebezem M, Beerthuyzen MM, Vaughan EE, de Vos WM, Kuipers OP (1997a) Controlled gene expression systems for lactic acid bacteria: transferable nisin-inducible expression cassettes for *Lactococcus*, *Leuconostoc*, and *Lactobacillus* spp. Appl Environ Microbiol 63:4581–4584
Kleerebezem M, Quadri LE, Kuipers OP, de Vos WM (1997b) Quorum sensing by peptide pheromones and two-component signal-transduction systems in Gram-positive bacteria. Mol Microbiol 24:895–904
Kolenbrander PE, Andersen RN, Moore LV (1989) Coaggregation of *Fusobacterium nucleatum*, *Selenomonas flueggei*, *Selenomonas infelix*, *Selenomonas noxia*, and *Selenomonas sputigena* with strains from 11 genera of oral bacteria. Infect Immun 57:3194–3203
Kolenbrander PE, Andersen RN, Blehert DS, Egland PG, Foster JS, Palmer RJ Jr (2002) Communication among oral bacteria. Microbiol Mol Biol Rev 66:486–505
Kreth J, Merritt J, Shi W, Qi F (2005) Co-ordinated bacteriocin production and competence development: a possible mechanism for taking up DNA from neighbouring species. Mol Microbiol 57:392–404
Kroes I, Lepp PW, Relman DA (1999) Bacterial diversity within the human subgingival crevice. Proc Natl Acad Sci USA 96:14547–14552
Kuipers OP, Beerthuyzen MM, de Ruyter PG, Luesink EJ, de Vos WM (1995) Autoregulation of nisin biosynthesis in *Lactococcus lactis* by signal transduction. J Biol Chem 270:27299–27304
Lazazzera BA (2000) Quorum sensing and starvation: signals for entry into stationary phase. Curr Opin Microbiol 3:177–182
Li YH, Hanna MN, Svensater G, Ellen RP, Cvitkovitch DG (2001) Cell density modulates acid adaptation in *Streptococcus mutans*: implications for survival in biofilms. J Bacteriol 183:6875–6884
Li YH, Lau PC, Tang N, Svensater G, Ellen RP, Cvitkovitch DG (2002) Novel two-component regulatory system involved in biofilm formation and acid resistance in *Streptococcus mutans*. J Bacteriol 184:6333–6342
McAuliffe O, Ross RP, Hill C (2001) Lantibiotics: structure, biosynthesis and mode of action. FEMS Microbiol Rev 25:285–308
Miller MB, Bassler BL (2001) Quorum sensing in bacteria. Annu Rev Microbiol 55:165–199
Nes IF, Holo H (2000) Class II antimicrobial peptides from lactic acid bacteria. Biopolymers 55:50–61
Nes IF, Diep DB, Havarstein LS, Brurberg MB, Eijsink V, Holo H (1996) Biosynthesis of bacteriocins in lactic acid bacteria. Antonie Van Leeuwenhoek 70:113–128
O'Toole GA, Kolter R (1998a) Flagellar and twitching motility are necessary for *Pseudomonas aeruginosa* biofilm development. Mol Microbiol 30:295–304
O'Toole GA, Kolter R (1998b) Initiation of biofilm formation in *Pseudomonas fluorescens* WCS365 proceeds via multiple, convergent signalling pathways: a genetic analysis. Mol Microbiol 28:449–461
Otto M (2001) *Staphylococcus aureus* and *Staphylococcus epidermidis* peptide pheromones produced by the accessory gene regulator agr system. Peptides 22:1603–1608
Palmer RJ Jr, Gordon SM, Cisar JO, Kolenbrander PE (2003) Coaggregation-mediated interactions of *Streptococci* and *Actinomyces* detected in initial human dental plaque. J Bacteriol 185:3400–3409
Pratt LA, Kolter R (1998) Genetic analysis of *Escherichia coli* biofilm formation: roles of flagella, motility, chemotaxis and type I pili. Mol Microbiol 30:285–293
Pratt LA, Kolter R (1999) Genetic analyses of bacterial biofilm formation. Curr Opin Microbiol 2:598–603

Prigent-Combaret C, Vidal O, Dorel C, Lejeune P (1999) Abiotic surface sensing and biofilm-dependent regulation of gene expression in *Escherichia coli*. J Bacteriol 181:5993–6002

Probert HM, Gibson GR (2002) Bacterial biofilms in the human gastrointestinal tract. Curr Issues Intest Microbiol 3:23–27

Qi F, Chen P, Caufield PW (2001) The group I strain of *Streptococcus mutans*, UA140, produces both the lantibiotic mutacin I and a nonlantibiotic bacteriocin, mutacin IV. Appl Environ Microbiol 67:15–21

Riley MA, Gordon DM (1992) A survey of Col plasmids in natural isolates of *Escherichia coli* and an investigation into the stability of Col-plasmid lineages. J Gen Microbiol 138(7):1345–1352

Riley MA, Gordon DM (1999) The ecological role of bacteriocins in bacterial competition. Trends Microbiol 7:129–133

Riley MA, Wertz JE (2002) Bacteriocins: evolution, ecology, and application. Annu Rev Microbiol 56:117–137

Scheie AA, Petersen FC (2004) The biofilm concept: Consequences for future prophylaxis of oral diseases? Crit Rev Oral Biol Med 15:4–12

Shapiro JA (1998) Thinking about bacterial populations as multicellular organisms. Annu Rev Microbiol 52:81–104

Smith JL, Fratamico PM, Novak JS (2004) Quorum sensing: a primer for food microbiologists. J Food Prot 67:1053–1070

Swift S, Downie JA, Whitehead NA, Barnard AM, Salmond GP, Williams P (2001) Quorum sensing as a population-density-dependent determinant of bacterial physiology. Adv Microb Physiol 45:199–270

Tait K, Sutherland IW (2002) Antagonistic interactions amongst bacteriocin-producing enteric bacteria in dual species biofilms. J Appl Microbiol 93:345–352

Testa MM, Ruiz de Valladares R, Benito de Cardenas IL (2003) Antagonistic interactions among *Fusobacterium nucleatum* and *Prevotella intermedia* with oral lactobacilli. Res Microbiol 154:669–675

Twomey D, Ross RP, Ryan M, Meaney B, Hill C (2002) Lantibiotics produced by lactic acid bacteria: structure, function and applications. Antonie Van Leeuwenhoek 82:165–185

Upton M, Tagg JR, Wescombe P, Jenkinson HF (2001) Intra- and interspecies signaling between *Streptococcus salivarius* and *Streptococcus pyogenes* mediated by SalA and SalA1 lantibiotic peptides. J Bacteriol 183:3931–3938

Wang BY, Kuramitsu HK (2005) Interactions between oral bacteria: inhibition of *Streptococcus mutans* bacteriocin production by *Streptococcus gordonii*. Appl Environ Microbiol 71:354–362

Watnick PI, Kolter R (1999) Steps in the development of a *Vibrio cholerae* El Tor biofilm. Mol Microbiol 34:586–595

Watnick P, Kolter R (2000) Biofilm, city of microbes. J Bacteriol 182:2675–2679

Whitehead NA, Barnard AM, Slater H, Simpson NJ, Salmond GP (2001) Quorum-sensing in Gram-negative bacteria. FEMS Microbiol Rev 25:365–404

Yother J, Trieu-Cuot P, Klaenhammer TR, De Vos WM (2002) Genetics of *Streptococci*, *Lactococci*, and *Enterococci*: review of the sixth international conference. J Bacteriol 184:6085–6092

Subject Index

A
activity spectrum: 95-101
allelopathy: 5, 17, 111, 113, 133, 134, 143
alveicin: 18, 20, 21, 29, 36-38
    -A: 18, 29, 36, 37, 38
    -B: 18, 29, 36, 37, 38
archaeocin: 3, 93, 94, 106, 107
autoregulation: 138

B
Bacillus: 2, 46, 55
    subtilis: 55, 57, 59, 137, 138
    megaterium: 79
bacteriocins:
    -class II: 48-74
        -aureocins: 67, 68, 70, 71
        -export mechanisms: 65, 66
        -pediocins: 64-66, 72, 82
        -plantaricins: 70, 73, 74
        -propionicins: 68, 71, 72
        -sakacins: 67, 71, 74
        -SLUSH peptides: 67, 70
        -streptocins STH_1/STH_2: 48, 70, 73
        -type IIa (pediocin-like): 64-66, 69
        -type IIb (multi-component): 66, 67, 69, 70
        -type IIc (miscellaneous): 67, 68, 71
        -ubericin A: 66, 69
    -class III (large proteins): 74-79
        -bacteriolysins (type IIIa): 74, 75, 77, 78
        -dysgalacticin: 79
        -lysostaphin: 74, 75, 77
        -non-lytic large proteins (type IIIb): 75, 76, 78, 79
        -zoocin A: 77, 78
    -class IV (cyclic peptides): 79-83
        -circularin A: 81, 82
        -enterocin A: 80-82
        -gassericin A: 82
        -reutericin 6: 82
        -uberolysin: 82, 83
    -Gram-positive: 45-92
        -bacteriocin-like inhibitory substance (BLIS): 49
        -cyclic peptides (Class IV): 79-83
        -detection methods: 48, 49, 51
        -classification: 50-52
        -lantibiotics: (Class I) 53-64
        -large heat-labile proteins (Class III): 74-79
        -megaplasmids: 60, 61
        -nomenclature: 49, 50
        -propionicins: 68, 71, 72
        -quorum-sensing peptides: 72-74
        -unmodified peptides (Class II): 64-72
    -multiple production of: 11-13, 16
    -resistance to: 15
biofilm: 135-137, 139-142
biotechnology: 93, 106, 143
bistability: 114, 115, 117, 121, 122, 129-130

C
cell-cell communication: 135, 137, 140, 142
cloacin: 20, 29, 35-38
    -DF13: 31, 34
    -683: 36, 38
    -647: 36
colicin: 9-15, 21-40, 112
    -5: 27, 28, 36
    -10: 28, 36
    -A: 9, 23, 28, 36-38, 40
    -B: 9, 12, 13, 15, 27, 28, 36, 40
    -D: 9, 10, 13, 24, 27-29, 35, 39, 40
    -E1: 9, 10, 12, 13, 16, 36

colicin (*Continued*)
   -E2: 9, 10, 12, 13, 33, 35
   -E3: 22, 25, 31, 34, 35, 37
   -E4: 24, 29, 31, 34, 35
   -E5: 24, 35
   -E6: 9, 24, 25, 27, 31, 34, 35
   -E7: 9, 12, 13, 24, 27, 31, 33, 35
   -E8: 22, 24, 31, 33, 35
   -E9: 22-24, 26, 29, 31, 33, 35
   -Ia: 9, 10, 12, 13, 15-17, 27, 28, 36, 38
   -Ib: 9, 13, 15, 27, 38
   -K: 9, 13, 23, 27-29, 33, 34, 36
   -M: 9, 12, 13, 15, 24
   -N: 36
   -nuclease: 22-24, 26, 27, 29, 31, 35
   -pore former: 22-24, 27, 29, 30, 32, 35, 36, 38
   -S4: 36
   -U: 28, 36
   -Y: 28, 36
competitive restraint (or survival of the weakest): 125-128
cost:
   -of bacteriocin production: 15, 113, 120
   -of resistance: 120, 125, 127
cross-induction: 128, 129

D
diversification: 20, 26, 27, 31
diversity (maintenance of): 112, 118, 119, 121-125, 128, 130
   -factors affecting: 7-13
      -in colicins: 7, 8, 11-13
      -in microcins: 10-13
   -of production: 9-13
      -in colicins: 9, 10, 12, 13
      -in microcins: 10-13

E
E. coli: 1, 2, 5-17, 20, 21, 28, 37, 38, 46, 66, 79, 90, 106, 112, 113, 117-121, 125, 130, 132-134, 142
Enterobacter: 20, 21, 28, 38
Enterobacteriaceae: 20, 21, 28
ESS: 120, 131-132
eucaryocin: 93, 94
evolution: 19-21, 23, 25-29, 31, 33, 35-37, 39

F
frequency of production: 7-12
   -colicins: 7, 11, 12
   -microcins: 11, 12

G
gastrointestinal tract: 64, 140-142

H
Hafnia: 8, 10
halocin:
   -A4: 95, 99, 100, 104, 106
   -C8: 95, 101, 104
   -G1: 95
   -H1: 95, 101, 104, 105
   -H2: 95
   -H3: 95
   -H4: 96, 101, 105
   -H6/H7: 93, 97, 103, 106
   -R1: 97, 99-103
   -S8: 97, 99, 100, 102

I
immunity: 21-27, 29-32, 35, 40

K
klebicin: 20, 21, 27-29, 31, 35-38, 40
   -B: 29, 31, 35, 37, 38, 40
   -C: 28, 29, 31, 35-37, 40
   -CCL: 31, 35, 36
   -D: 27-29, 35, 37, 38, 40
Klebsiella: 20, 21, 28, 29, 38

L
Lactobacillus: 79
   acidophilus: 65, 82
   casei: 67, 78
   curvatus: 142
   gasseri: 80, 82
   helveticus: 75, 79
   johnsonii: 67
   lactis: 53-57, 62, 67, 69, 79, 138
   plantarum: 70, 73
   reuteri: 80, 82
   sakei: 63, 71, 74
   salivarius: 67
lantibiotics: 2, 45, 49-51, 53-64, 66, 73, 82, 100, 138
   -cinnamycin: 55, 61, 62

Subject Index

-enterococcal cytolysin: 55, 63, 64
-lacticin 3147: 55, 62
  -mersacidin: 55, 61
  -mutacin: 2, 49-51, 56, 67, 71, 73, 141
-nisin: 2, 47, 50, 51, 53, 54, 56-59, 61, 65, 72, 82, 92, 100, 138
-salivaricin A: 54, 59, 60
-streptococcin A-FF22 (SA-FF22): 54, 59
-subtype AI (nisin-like): 53, 54, 56-58
-subtype AII (SA-FF22-like): 54, 58-61
-type B (globular): 55, 61, 62
-type C (multi-component): 55, 62-64

M
marcescin: 20, 21, 29, 36-38
  -A: 29, 36, 38
metapopulation: 119
microcins: 8, 11-17, 20, 46, 66
  -B17: 11, 13
  -C7: 11, 13
  -H47: 11-13, 16
  -J25: 11-13
  -L: 11, 13
  -M: 11-13, 16
  -V: 10-13, 16, 17
model:
  -mathematical: 7-8, 15, 113-120, 122, 130-132, 140
  -simulation: 118, 120, 122-130
  -community: 112, 114, 117, 118, 121, 125, 128, 130
mutation: 25-27, 29, 60, 106, 119, 125-126, 128

N
niche construction (or ecosystem engineering): 129-131

O
oral cavity: 47, 60, 64, 140

P
polymorphism: 25
Pseudomonas: 20, 21
pyocins: 2, 20-25, 28, 29, 31, 35, 37-40
  -AP41: 31, 33, 35, 40
  -F-type: 21, 40
  -F2: 22
  -R-type: 21, 40
  -R2: 20, 22
  -S-type: 21, 22, 24, 29, 37
  -S1: 24, 28, 33, 35, 38, 40
  -S2: 24, 25, 31, 33, 35, 38, 40
  -S3: 25, 35, 40
  -S5: 22, 24

Q
quorum sensing: 63, 72, 73, 135-139

R
recombination: 25, 27-29, 38
regulation: 2, 37-38, 56, 59, 74, 102, 136, 138
resistance: 6, 10, 11, 16, 47, 53, 119, 120, 125-128
rock-paper-scissors (or non-transitivity): 14, 99, 119-128, 130

S
selection:
  -positive: 25, 26
  -diversifying: 25-29
sequence: 2, 19, 21, 22, 25, 30, 35, 37-41, 49, 50, 52, 53, 57-59, 67, 68, 72, 77, 79, 81, 82, 94, 98, 100, 102-105, 123
Serratia: 20-22, 28, 36-38
signal transduction: 54, 68, 70, 72-74, 135, 136, 138, 139
spatial structure: 117, 118, 120-130
Staphylococcus: 2, 46, 55, 61,
  aureus: 47, 49, 55, 61, 67-71, 77, 78, 137
  capitis: 77
  epidermidis: 77
  lugdunensis: 67, 70
  simulans: 74, 75, 79
Streptococcus: 2, 57, 59, 66, 77, 80
  agalactiae: 60
  constellatus: 78
  dygalactiae: 60, 75, 79
  equi: 75, 77
  gordonii: 49, 70, 73, 78, 137, 141
  macedonicus: 49
  milleri: 75, 78
  mutans: 49-51, 55, 62, 67, 71, 73, 137, 141
  pneumoniae: 73, 137
  pyogenes: 49, 54, 57-60, 76, 79, 138, 141

Streptococcus (*Continued*)
  rattus: 63, 68
  salivarius: 49-51, 54, 59-61, 67, 138, 140, 141
  thermophilus: 67, 69
  uberis: 49, 51, 54, 57, 66, 69, 80, 82, 83
sulfolobicin: 3, 93, 94, 106, 107